Sustainable Development and Construction in Asia

This book illuminates fundamental knowledge and provides a comprehensive guide for those seeking to understand the intricacies of sustainable development and construction in Asia. Each chapter is dedicated to a specific aspect of sustainable construction, offering in-depth analysis, case studies, and practical insights.

The main characteristic of this book is a comprehensive exploration and integration of research, practices, and innovations, emphasizing the importance of academic research and practical implementations in driving sustainable construction. The advanced project management methods towards sustainable construction are explained, including life cycle management, sustainable procurement, risk management, lean construction, and integrated project delivery. Smart construction technologies are illustrated, for example, 3D printing, BIM (building information modelling), industrialized, prefabricated, and digital technologies like eXtended reality, machine learning, digital twins, big data, blockchain, Internet of Things, and cloud construction technologies. Particularly, the relationships and differences among off-site, industrialized, prefabricated, modular, panelized, and hybrid construction are displayed. Finally, practical on-site construction techniques and valuable sustainable construction materials are demonstrated in shaping sustainable practices.

This book will be of interest to practitioners, researchers, and consultants in the general field of sustainable development in construction.

Larry Xiancun Hu has an extensive background in the construction industry, encompassing research, education, and practice in both Australia and China. He is currently an assistant professor in the building and construction management discipline at the University of Canberra. His research focuses primarily on sustainable construction practices, smart and digital construction technologies, and advanced construction project management. His work aims to promote cost reduction, value creation, carbon reduction, and sustainable productivity in the construction industry.

Routledge Research in Sustainable Planning and Development in Asia

Series Editor: Richard Hu

Data-centric Regenerative Built Environment
Big Data for Sustainable Regeneration
Saeed Banihashemi and Sepideh Zarepour Sohi

Disaster Resilience and Sustainability
Japan's Urban Development and Social Capital
Hitomi Nakanishi

Megaregional China
Richard Hu

Building Urban Resilience
Singapore's Policy Response to Covid-19
Junjie Woo and Debbie R. Loo

Urban Renewal in Central Seoul
Planning Paradigm Shifts
Hyung Min Kim

Pursuing Sustainable Urban Development in North Korea
Edited by Pavel P. Em

Data-Driven Smart Community Design
Strategies for Fostering Inclusive and Resilient Neighbourhood
Edited by Keng Hua Chong

Sustainable Development and Construction in Asia
Larry Xiancun Hu

For more information about this series, please visit: www.routledge.com/Routledge-Research-in-Sustainable-Planning-and-Development-in-Asia/book-series/RRSPDA

Sustainable Development and Construction in Asia

Larry Xiancun Hu

LONDON AND NEW YORK

First published 2025
by Routledge
4 Park Square, Milton Park, Abingdon, Oxon, OX14 4RN

and by Routledge
605 Third Avenue, New York, NY 10158

Routledge is an imprint of the Taylor & Francis Group, an informa business

British Library Cataloguing-in-Publication Data
A catalogue record for this book is available from the British Library

ISBN: 978-1-032-06521-2 (hbk)
ISBN: 978-1-032-06523-6 (pbk)
ISBN: 978-1-003-20266-0 (ebk)

DOI: 10.4324/9781003202660

Typeset in Galliard
by Apex CoVantage, LLC

Contents

Preface

The construction industry stands at a pivotal moment, balancing economic performance with environmental protection under the principles of sustainability. With Asia experiencing unprecedented growth, the urgency for sustainable construction practices has never been more critical. This book, "Sustainable Development and Construction in Asia", provides a holistic view of how sustainable principles are being integrated into construction projects across Asia.

This book illuminates fundamental knowledge and provides a comprehensive guide for those seeking to understand the intricacies of sustainable development and construction in Asia. Each chapter is dedicated to a specific aspect of sustainable construction, offering in-depth analysis, case studies, and practical insights. Chapter 1 introduces the concept and significance of sustainable development and construction in the Asian context. Chapter 2 explores the circumstances, barriers, and solutions of sustainable construction practices in Asia through case reviews. Subsequent chapters delve into advanced project management methods, smart construction technologies, on-site construction techniques, and sustainable construction materials that have been and can further be developed to promote sustainable construction practices and performance.

The main characteristic of this book is a comprehensive exploration and integration of research, practices, and innovations, emphasizing the importance of academic research and practical implementations in driving sustainable construction. For, the differences in concept and scope among green buildings, green construction, and sustainable construction are outlined. The advanced project management methods towards sustainable construction are presented, including life cycle management, sustainable procurement, risk management, lean construction, and integrated project delivery methods. Smart construction technologies are illustrated, such as 3D printing, BIM (building information modelling), industrialized, prefabricated, and digital technologies like eXtended reality, machine learning, digital twins, big data, blockchain, Internet of Things, and cloud construction technologies. Particularly, the relationships and differences among off-site, industrialized, prefabricated, modular, panelized, and hybrid construction are displayed. Finally, practical on-site construction techniques and valuable sustainable construction materials are demonstrated in shaping sustainable practices.

With 25 years of experience in the construction field, this marks my debut book dedicated to shaping a sustainable future in construction. I trust that this book will serve as a valuable resource for scholars and students in construction management studies. It is my aspiration that policymakers, researchers, industry professionals, and sustainability practitioners within the construction sector will find fresh perspectives on sustainable construction practices within these pages.

I express my sincere gratitude to all the references and contributors whose wealth of knowledge and exploration have enriched the content of this book. I am deeply thankful to the University of Canberra and the NSW Government Environmental Trust for their unwavering commitment to sustainability and supporting education and research in this vital area. A special mention of appreciation goes to Prof Richard Hu and Dr Aifang Wei from the University of Canberra (Australia) and Dr Ruixue Zhu from Shandong Jianzhu University (China) for their invaluable recommendations and insights.

Thank you for embarking on this journey with me. May it offer valuable insights and practical solutions that can be implemented in real-world projects, contributing to a more sustainable future for the construction industry and a more harmonious future for the Earth.

Larry Xiancun Hu
Canberra, Australia
July 2024

Abbreviation

AI	Artificial intelligence
AIA	American Institute of Architects
BIM	Building information modelling
CO_2	Carbon dioxide
EMP	Energy management practices
ERs	Environmental risks
GDP	Gross domestic product
GHG	Greenhouse gas
GNP	Gross national product
HCP	Highway construction projects
HVAC	Heating, ventilation, and air conditioning
IoT	Internet of Things
IPD	Integrated project delivery
JIT	Just-in-time
LCA	Life cycle assessment
LCCM	Life cycle cost management
LCM	Life cycle management
LEED	Leadership in Energy & Environmental Design Building Rating System
ML	Machine learning
MR	Mixed reality
OECD	Organization for Economic Co-operation and Development
PDC	Plan-Do-Check-Act
R&D	Research and development
SDG	Sustainable Development Goals
SP	Sustainable procurement
SWOT	Strengths, weaknesses, opportunities, and threats
TQM	Total quality management
UAE	United Arab Emirates
UN	United Nations
UNEP	United Nations Environment Programme
VR	Virtual reality
WHS	Work health and safety
XR	eXtended reality

1 Sustainable development in the construction industry

Introduction

1.1 Sustainable development

Development has brought environmental problems which have inevitably restricted and affected its own development. This situation stems from a combination of factors, including population explosion, resource depletion, urban expansion, and regional challenges. In order to measure and alleviate the environmental problems that threaten the survival and further development of humans, the United Nations (UN) convened many crucial conferences (see Table 1.1), including the UN Conference on the Human Environment in 1972, *Our Common Future* (also known as the Brundtland Report) in 1987, and UN Conference on Environment and Development in 1992, which are regarded as the important milestones in the formation and development of the concept of "sustainable". Table 1.1 displays the milestones for sustainable development. Particularly, in 2022, the UN General Assembly convened the international meeting of Stockholm+50 saying "a healthy planet for the prosperity of all—our responsibility, our opportunity". In summary, the idea of sustainable development is formed on the basis of people's deepening understanding of the connections among the population, economy, resources, and environment. The way of human sustainable development has never been stopped, and it will continue to make efforts in guiding the whole world towards sustainable development.

The concept of sustainable development has been widely discussed in the world and has formed a relatively complete theoretical system. Sustainable development involves two concepts: "development" and "sustainability".

The term "development" was originally defined by economists and has been expanded from a simple economic concept to concern with human needs and social progress. At first, the goal of economic development is to increase output and profits; therefore, the pursuit of gross national product (GNP) growth has become the goal and motivation of national economic development. This kind of development destroys resources and the environment in pursuit of maximum economic benefits. The serious deterioration of the global environment has made human beings realize the irrationality of such development. Growth that focuses solely on increasing output and improving

DOI: 10.4324/9781003202660-1

Table 1.1 Milestones for sustainable development

Year	*Milestone*
1972	Declaration of the UN Conference on the Human Environment, Stockholm
1987	UN Brundtland Commission; World Commission on Environment and Development, *Our Common Future*
1992	Rio Declaration on Environment and Development, in the Report of the UN Conference on Environment and Development, Rio de Janeiro
1997	19th Special Session of the General Assembly to Review and Appraise the Implementation of Agenda 21, New York
2000	Millennium Development Goals Report 2015, New York
2002	World Summit on Sustainable Development, Johannesburg
2005	World Summit, The Millennium Development Goals: Five years later, New York
2008	High-level meeting on the Millennium Development Goals, New York
2010	Millennium Development Goals Summit, A Global Strategy for Women's and Children's Health, New York
2012	UN Conference on Sustainable Development (Sustainable Development Goals, SDGs), Rio de Janeiro
2013	President of the General Assembly's Special Event towards Achieving the Millennium Development Goals, New York
2014	Conference Diplomacy and the World's Growing Commitment to Sustainable Development
2015	UN Summit on Sustainable Development, Changing Our World, New York
	The Paris Agreement reduces global temperature rise and avoids the impacts of climate change, Paris
2022	Stockholm+50, 2030 Agenda and the Sustainable Development Goals

the GNP is not sustainable development; it merely represents the quantitative expansion of a country's economic scale. Therefore, development should be a comprehensive performance of a country's economic, social, environmental, cultural, and structural conditions. It involves improving the quality of life of individuals and communities through various means, such as education, health, and infrastructure development.

It is therefore generally accepted that development is steady and sustained economic growth without environmental harm, and that development should be primarily an improvement in the progress of society including economic development, social development, environmental development and ecological development. In summary, development refers to the continuous growth of economic, environmental, and social aspects.

The concept of "sustainability" originates from the use of renewable resources such as forests and fisheries. This concept then extends to the ecosystem, integrating with economy, society, environment, and growth to form the concepts of *sustainable economy*, *sustainable society*, *sustainable environment*, and *sustainable growth*, respectively. The transformation of development mode refers to the gradual change from resource-consuming development to resource-saving development. That is, relying on scientific and technological progress; saving resources and energy; reducing waste discharge; implementing clean production; and establishing the coordination of economy, society, resources, and environment. Therefore, sustainability can be integrated with a certain object to combine into a composite concept, such as ecological sustainability. When sustainability is integrated with the development of an object, it constitutes the concept of *sustainable development of an object*, such as environmental sustainable development, social sustainable development, and economic sustainable development.

The World Commission on Environment and Development (1987) develops the definition that "sustainable development is development that meets the needs of the present without compromising the ability of future generations to meet their own needs". Therefore, the definition of sustainable development has different connotations: when focusing on the environment, sustainable development can be defined as protecting and strengthening the productive and regenerative capacity of environmental systems; when focusing on social attributes, it can be defined as improving the quality of human life while maintaining survival and development within the capacity to sustain the ecosystem; when focusing on economic attributes, it can be described as that the use of economic resources should not reduce future real income, environmental quality, and social development of the world's natural resources; and when focusing on technological attributes, it can be seen as the use of cleaner and more efficient technologies to minimize the consumption of energy and resources. Therefore, sustainable development naturally involves balancing environmental, social, and economic factors to create a sustainable future for human society.

"The world is faced with challenges in all three dimensions of sustainable development—economic, social, and environmental" (United Nations, 2013). For instance, global warming, one of the most serious environmental issues, is now a critical challenge for sustainable global development, as it has led to various developmental problems such as long periods of drought, rising sea levels coupled with coastal flooding, the possible spread of infectious diseases, and effects on atmospheric and ocean circulation that impact rainfall and wind patterns. Therefore, in 2015, the United Nations developed 17 Sustainable Development Goals (SDGs), also known as the Global Goals, which put forward a set of universal goals to meet the urgent environmental, political, and economic challenges in order to call to action to end poverty, protect the planet, and ensure peace and prosperity of all people in the world by 2030. The 17 SDGs aim to address the development from social, economic, and environmental dimensions and recognize that action in one affects success for others. The development in three dimensions must be balanced and shift to a sustainable development path from 2015 to 2030. Therefore, they cover the issues that affect us all and affirm international commitment to build a more sustainable, safer, more prosperous planet for all humanity.

1.2 Why is sustainable development important for the construction industry?

With the continuous increase of population and the acceleration of urbanization, the construction industry plays an extremely important role in the global economy. However, the underperformance of the construction industry can be demonstrated in various aspects, such as consuming more resources (e.g. materials, energy, labour, land) than needed, producing fewer contributions (e.g. value-added, profit) than expected, and generating more waste and pollution (e.g. construction waste, greenhouse gas (GHG) emissions) than desired. These problems can be summarized as economic problems, environmental problems, and social problems.

1.2.1 Environmental problems

Environmental problems have become a serious global issue in the building and construction industry, including energy consumption, pollution generation, land use, and other issues. The building and construction sector in the world contributes 36% and 37% to global final energy consumption and energy-related greenhouse gas (GHG) emissions, respectively (United Nations Environment Programme, UNEP, 2021). A construction project is a large consumer of resources that include energy, water resources, land resources, building materials, and various mineral resources needed in the production process. Furthermore, it is evident that the construction industry traditionally has been slow to adopt sustainable practices, such as green building design, energy-efficient technologies, and waste reduction strategies, leading to increased environmental impact.

Construction activities have negative impacts on natural habitats and the natural behaviour of wildlife. As an illustration, construction operations have a substantial impact on both land-based and aquatic plant and animal life by involving in activities like clearing indigenous vegetation; carrying out tasks near and within water bodies; creating noise, vibrations, and light; disrupting soils; and utilizing chemicals and fuels (which may lead to spills). According to the fact sheet from the World Wildlife Fund in 2017, the problem of impacting habitats and wildlife from human activities in Australia is exceptionally severe, such that more mammal species have been lost more than in all other continents combined over the past two centuries. More importantly, human error activities can also impact the environment negatively during construction. Due to the complexity of building and the long-term process of building construction, human errors are not uncommon in different building components and construction stages due to a wide range of reasons, which can not only contribute to more energy and material consumption but also cause more GHG emissions and waste. Therefore, the negative effects need to be controlled to achieve environmental sustainability in the construction industry.

1.2.2 Economic issues

The construction industry plays an important role in the economic market. The construction industry is one of the significant sectors accounting for a considerable proportion of the national gross domestic product (GDP) and is the largest industrial employer in most countries. Moreover, the acceleration of urbanization has further promoted economic development and stimulated the continuous expansion of the construction industry. However, a series of problems has also been caused by the rapid development of the construction industry. For instance, the current unsustainable, low-profitability, and underperformance conditions of the construction industry have been subject to wide criticism. More importantly, a green economy is a new economic subdiscipline, which is based on the traditional industrial economy and aims to promote the balance of development between the economy and the environment. How to achieve sustainable development in a green economy is a challenge for the construction industry.

From a micro perspective, another primary economic issue in the construction industry is hard-to-achieve construction project management goals in numerous projects, such as wasting material, over budget, schedule delay, and scope changes. The key reason for this issue could be unclear project objectives which will result in confusion, miscommunication, scope creep, conflicts, and contract changes. The construction project management practices and performance significantly impact the achievement of project management goals, where factors such as inadequate planning and scheduling, ineffective communication and collaboration, poor risk management, lack of efficient technologies and skills, and insufficient stakeholder engagement and satisfaction can all influence outcomes.

1.2.3 Social issues

There are many kinds of social issues in the construction sector. The most important issues could consist of work health and safety (WHS) issues, workforce and skills shortage, and lack of advanced technology adoption.

WHS pertains to managing risks to the health and safety of all individuals present at construction sites. Construction sites that are close to residential areas can seriously influence residents' normal work and life. In addition, on-site workers are exposed to noise hazards due to the construction machinery working in open places with no shelter around them. For example, among the hazards caused by air pollutants, suspended particulate matter can be inhaled and directly deposited in the lungs, affecting lung function and causing various respiratory diseases. Therefore, in the harsh construction environment, how to ensure the health and safety of construction workers is an important issue. It is impossible to change harsh weather conditions or move construction projects from the wilderness to cities, but to protect the health of workers and to better protect the environment are what a sustainable construction means.

Workforce and/or skills shortages have become a primary issue in the construction sectors in most countries for the past few years, which have significantly affected construction progress, costs, efficiency, and overall performance. The sector is confronting difficulty in attracting new, younger talent to enter and remain, particularly the female talent. The reasons could be long working hours, low payment, low safety rates, and harsh site conditions. Particularly, along with the requirements of sustainable development and Industry 4.0, construction employees find it difficult to fill vacancies for an occupation or specialized skills such as the technologies of building information modelling (BIM), drones, artificial intelligence (AI), and the Internet of Things (IoT).

Compared to other sectors, the construction sector has been criticized for slowly adopting advanced technologies, which will negatively promote construction efficiency and sustainable development. The lack of a skilled workforce is one of the reasons that contribute to the slow adoption. More importantly, the construction processes are highly fragmented and involve many stakeholders and participants including owners, architects, engineers, consultants, contractors, subcontractors, and suppliers. "Information islands" commonly exist during the construction process among these participants. It is very hard to communicate, coordinate, and implement new technologies for all participants across the various stages of a project. The other factors that hinder the use of new technologies include cost and time-consuming considerations, economic fluctuations and uncertainties, no obvious benefits for one project particularly for small projects, and the comfort of sticking to traditional construction methods.

As a consequence, sustainability is particularly a major concern for governments, clients, and stakeholder groups in the construction industry. Following increasing emphasis on sustainability, the construction sector must become more accountable for sustainable development issues.

1.3 Sustainable development requirements in the Asian construction market

1.3.1 The pressure of climate changes and energy consumption

Climate change poses a significant threat to sustainable global development, manifesting in extended periods of drought, elevated sea levels resulting in coastal flooding, potential proliferation of infectious diseases, and disruptions to atmospheric and oceanic circulation. Particularly, extreme weather events, such as the most serious natural disasters, can severely damage living conditions and asset value. For example, the coastal areas of Queensland and New South Wales in eastern Australia experienced one of Australia's worst flood disasters with a series of extreme floods following days of rain from late February to early March 2022. The flooding resulted in 22 deaths, and more than 25,000 homes were damaged, with an estimated total value of

AU$1.77 billion in insurance claims. Furthermore, heavy rainfall in the area had significantly impacted water quality and hit the safety of the state's rivers and beaches. Consequently, climate change, particularly extreme weather events, will increasingly affect the built environment and put pressure on society as a whole.

The increases in energy consumption and GHG emissions are the primary reasons for climate changes, which are largely due to human activity, such as the use of fossil fuels (coal, oil, and natural gas), agriculture, building and construction, and other land use activities. According to the report *GlobalABC Regional Roadmap for Buildings and Construction in Asia 2020–2050* (GlobalABC/IEA/UNEP, 2020), the buildings and construction sector plays a major part in achieving the vision of the global temperature increase in the Paris Agreement, as the sector accounted for 36% of final energy use and 39% of emissions of carbon dioxide (CO_2) from energy use and industrial processes. More significantly, in order to upgrade and secure people's life quality, the high living standards in industrialized countries and the progress of living conditions in developing and newly industrialized countries are continuously promoting the demand for building and infrastructure construction, which substantially requires an increasing consumption of energy and other natural resources. Therefore, it is a long-term challenge for the development of the building and construction sector and the society as a whole.

In Asia, in 2018, the buildings sector in the Association of Southeast Asian Nations (ASEAN), the Republic of China (hereafter referred to as "China") and India accounted for 27% of the region's final energy use and 24% of energy-related CO_2 emissions in the world, excluding emissions from manufacturing building materials (IEA, 2019). Since 2010, growth in regional energy demand has been driven by a 7% increase in population and a 70% increase in wealth (GDP) which, in turn, has increased the demand for floor area and energy-consuming services. Continued growth in floor area is expected in the region which could represent nearly half of the new construction in the world by 2040 (GlobalABC/IEA/UNEP, 2020). Moreover, the use of fuels in buildings varies considerably, for example, electricity is most consumed in China while biomass utilization constitutes a significant portion in India. Anyway, most construction sectors in Asian countries require a decrease in the consumption of natural sources and the emission of environmentally harmful substances.

1.3.2 Social requirements

In the Asian construction market, previous studies have focused extensively on the economic and environmental aspects of sustainability. However, the social dimension has garnered comparatively less attention due to factors such as insufficient theoretical foundations and practical support, difficulties in measuring social sustainability performance and progress, and challenges in directly observing and experiencing its outcomes. Nevertheless, promoting social sustainability holds significant value for the construction industry for the following reasons.

First, construction products are a key element in determining the quality of the environment in which society lives and works. For example, improving the building quality and services could benefit the living environment and human health and comfort, which involves building performance during the whole life cycle through design, construction, operation, and maintenance. Particularly, social infrastructure supports access to social services for communities and the public and provides societies with their physical and functional environments, such as culture-related, health-related, education-related, and government-related facilities and spaces. Old and historical buildings are also a symbol of cultural identity and heritage, which can improve the well-being of communities and strengthen economic sustainable development, such as through the tourism sector.

Second, the construction sector is one of the single largest industrial sectors which provide employment and value. As a labour-intensive industry, it provides a relatively fair environment for various persons who could have opportunities to work in the construction industry without the high constraints of knowledge, skills, and abilities. Besides, the construction industry is an important material production department of the national economy. The construction products represent a considerable segment of the fixed assets of individuals, organizations and nations. Therefore, it could significantly support poverty reduction among individuals through work and social services, and between the poorest countries and rich countries through the export of labour and materials. For example, according to the report *Migrant Work & Employment in the Construction Sector* (Buckley et al., 2016), the construction industry in the United Arab Emirates (UAE) is almost entirely made up of foreign workers; an estimated 700,000 construction migrants had entered the UAE by the mid-2000s; and the workers were mainly classified as low- and semi-skilled from India, Pakistan, Bangladesh, Nepal, Sri Lanka, and the Philippines. Moreover, the construction migrants could help keep labour costs down and sustain the construction market boom.

Third, the construction industry is one of the industries which can be easily and comprehensively influenced by various social factors. First, the construction scale is significantly influenced by government investment, particularly in developing countries. For instance, China allocated RMB 3.65 trillion (US$573 billion) for infrastructure projects in 2021, in which 50% was invested in infrastructure, municipal administration, and industrial park infrastructure sectors, and 30% was used on social projects such as affordable housing (Huld, 2022). Second, the regulations, codes, standards, and policies have seriously impacted on construction process and products. For example, the regulations, codes, standards, and policies relevant to building design and construction will directly influence building quality and cost, which will further influence the improvement of people's lives and the development of the whole national economy. An expansionary or loose monetary policy will speed up the growth rate of the construction industry. In contrast, a tight or contractionary monetary policy will slow down the growth rate. Third, local contents and communities have various influences on the projects during the whole construction process from project initiation, planning, design, construction, and delivery. One of the sustainability requirements shall be that the construction projects shall not produce damage to local communities and environments.

Finally, ethical, fair work, equity, and access to human resources are essential to supporting employees throughout their employment and industry to voice concerns, prevent workplace bullying, and sustain strong and innovative strategies in relation to corporate social responsibility. Furthermore, all participants in the construction sector could get the correct payment, remuneration, and benefits. For instance, construction employees will have correct payment, workers' compensation, superannuation, and fair work conditions. All construction stakeholders, particularly governments, shall ensure these key aspects are as per the minimum legislative requirements enforced. When these requirements are not met, it greatly affects human well-being and industry growth, particularly through the sustainable development viewpoint. It is integral that the construction industry as a whole meets these social requirements to protect participants in the industry.

1.3.3 Economic pressures

The construction sector is a significant component of the economic landscape and is a significant contributor to the quality of life. The construction industry is one of the chief sectors which accounts for a considerable proportion of national GDP and is the largest industrial employer in most countries, particularly in developing countries. Construction, building materials, and associated professional activities account for a significant percentage of the GDP. For example, according to the report from *Construction World*, India's premier and largest circulated construction business magazine, the compound annual growth rate of the Indian construction industry has increased from 6.3% during 1993–2000 to 9.17% during 2000–2005 and then improved to 10.31% during 2005–2010. Employment in the Indian construction sector has enhanced from 8.5 million in 1993–1994 to 17.50 million in 1999–2000, and then further increased to 26 million during 2004–2005 and 44.10 million during 2009–2010. Therefore, the construction industry not only is a major source of economic development in the national economy but also plays a significant role in advancing socio-economic and sustainable development for the whole country.

The significance of the construction industry is embodied not only in its contributions but also in its importance for promoting economic growth for other sectors, probably due to the fact that the construction sector is strongly linked to other economic activities through the pull and push effects. For example, building construction will consume large amounts of different materials, where these materials, being intermediate inputs from supply industries, are delivered to the housing construction industry. Accordingly, there is an economic pull effect towards these supplier industries. On the other hand, there can be an economic push effect from the construction industry towards downstream industries when construction products such as real estate, infrastructure, and industrial structures are demanded by other industries. Importantly, strong linkages between the construction industry and other industries are reflected in income growth, activity generation, employment increase, and the stimulation of demand.

However, the role of the construction in the economy is dynamic and exhibits an inverted U-shaped curve, named the "Bon Curve" as well, based on the study of Bon (1992) (see Figure 1.1). The Bon Curves present the changing role of the construction sector at three stages of economic development. In the first stage, the role increases along with the economic development, for example, when the infrastructure and industrial buildings are built during the Industrial Revolution processes and the residual buildings are required to improve the living environment, particularly for low-income countries. Accordingly, the construction scale, value added, and GDP share in nations significantly improve, for instance, in the Indian construction industry. In the second stage, the role coincides with the smooth peak of the Bon curve, when the construction market is in maturity. In the third stage, the role and share of construction in the overall economy subsequently decline with the level of economic development. The reasons could be that infrastructure becomes more complete and developed, and housing shortages are eliminated when the quality of living is improved. For example, the pull and push linkages between the Australian construction industry and the overall economy weakened from 2000 to 2014 due to social, economic, and technological developments in Australia, with annual decreases of approximately 0.8% and 0.5%, respectively. In Asia, where most countries are developing, the construction industries have significant potential for improvement alongside national economic

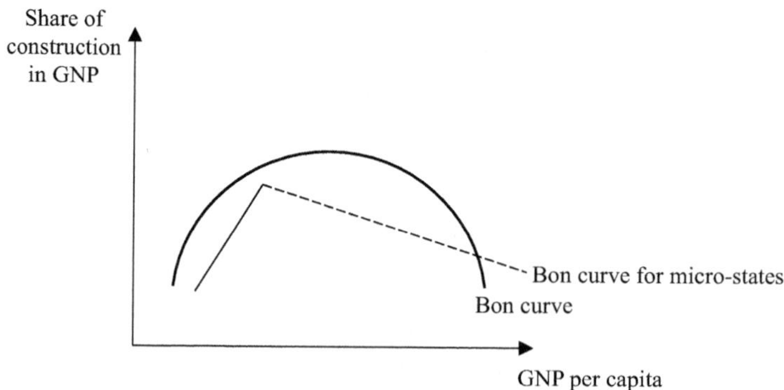

Figure 1.1 The inverted U-shaped "Bon Curve" of the construction industry on micro-states.

development. However, it is also crucial to explore how to promote the sustainable development of construction industries, especially in addressing the declining trends in the third stage after achieving a peak in the second stage.

1.3.4 Human well-being

Social sustainability in construction involves addressing the needs of people throughout all stages of the construction process, ensuring high customer satisfaction, and collaborating closely with clients, suppliers, employers, and the local community. Sustainable construction will minimize resource depletion and environmental degradation while creating a healthy built environment. Therefore, human factor shall be taken into consideration when improving social sustainability. Actually, sustainability is about people, consisting of currently living generations and future generations. The improvement of economic and environmental sustainability will provide support for people's living conditions and environment. Therefore, WHS is one aspect of social sustainability. Particularly, as the construction sector is one of the sectors with the most workers globally, the WHS issue is a significant and common development challenge in most construction industries in the world. In Indonesia, the construction safety and health accident rate is highest across all sectors and is growing annually by 11% from 2013 to 2018. In contrast, Australia decreased the average incident rate of construction safety and health by 2% from 2013 to 2018. Currently, people are more aware of work safety and health. Consequently, improving WHS performance should be a crucial aspect of sustainable development in the construction industry.

The well-being of individuals and communities should be considered in social sustainability, such as psychological and physiological needs. Construction encompasses various social factors that directly impact the well-being of humans. For example, sustainable design involves considering how a building will be designed to facilitate health and physical comfort throughout its entire life cycle. The design not only influences the construction process to ensure that construction is completed with quality, within budget, and on time but also impacts the building's operation, maintenance, and demolition, promoting occupant usability, operational efficiency, community interaction, and overall living quality.

Integrating WHS into sustainable construction strategies could improve social sustainability and truly achieve sustainable development in the construction industry. No countries, industries, or organizations can be sustainable without prioritizing the safety, health, and welfare of their people. In construction projects, improving WHS can lead to a decrease in the number of illnesses, injuries, and fatalities, as well as an increase in construction quality and productivity. Furthermore, this can reduce costs associated with workers' compensation, training, and recruitment, and enhance the satisfaction and reputation of construction companies. Overall, embracing WHS as a cornerstone of sustainable construction is good for workers, stakeholders, and businesses, and good for people to thrive.

The COVID-19 pandemic has severely impacted the development of the construction market and output value. Particularly, the construction industry in most Asian countries has suffered deep contractions in 2020. For example, in 2020, the construction output value in Singapore, the Philippines, and Malaysia decreased by 36%, 30%, and 20%, respectively, according to the 2021 Global Status Report for Buildings and Construction by the UNEP. Additionally, China's Ministry of Foreign Affairs reported in June 2020 that 30% to 40% of Belt Road Initiative projects were affected by the pandemic, with another 20% experiencing significant delays. In contrast, the Indonesian construction industry saw a 3.3% decrease in output value. Similarly, Qatar and Saudi Arabia in the Middle East experienced contraction rates of 4.2% and 0.4% in their construction outputs, respectively, demonstrating slightly greater stability compared to the southeast Asian countries. Accordingly, due to various local market conditions, development strategies, and pandemic response policies, different countries exhibited different contraction rates as a result of the pandemic's impact on the construction industry.

1.4 What is sustainable construction?

Sustainable development has been gaining momentum across the construction industry all over the world within the past 30 years. In June 1992, the *United Nations Conference on Environment and Development* (UNCED) issued programmatic documents such as the *Rio Declaration on Environment and Development* and *Agenda 21*, proposing that the development of human society should take the road of sustainable development to harmonious coexistence. A new strategy for the development of human society at this conference was identified, which promoted the design principles of building 3Rs, namely, reducing the use of non-renewable energy and resources, recycling building components or products, and reusing certain constituent materials in restoration. In 1994, the *First International Conference on Sustainable Construction* put forward the basic concept of "sustainable construction", that is, integrating sustainable development into the whole life cycle of design, construction, operation and maintenance, renovation, and demolition to maximize the effective use of non-renewable resources to reduce pollutant discharge and their impact on human health, to create a green environment which is conducive to human survival and development. In June 1996, the *Second United Nations Conference on Human Settlements (Habitat II)* addressed two themes of equal global importance: Human beings are at the centre of concerns for sustainable development, including "adequate shelter for all" and "development in an urbanizing world". In October 2000, the International Conference on Sustainable Building 2000 was held in Maastricht, Netherlands, and proposed to promote the development of building environmental assessment methods and technologies.

In March 2005, a sustainable development conference was held in Japan to further discuss and improve the green building assessment tool.

With the development of a green economy, more attention has been paid to sustainable development in the construction sector. Green economy in the sector emphasizes environmental protection by reducing energy consumption in buildings and by providing healthy, comfortable, and efficient space for human beings to realize the harmonious coexistence between buildings and nature. Generally, the realization of sustainable construction is of great significance in three main aspects: (1) in terms of economic benefit, it can effectively reduce the operation cost of construction projects and increase the value of construction; (2) in terms of social benefits, sustainable construction can improve the quality of people's work and life; (3) in terms of environmental benefits, sustainable construction can reduce environmental pollution, improve environmental quality, reduce the consumption of non-renewable, resources and improve the utilization efficiency of resources. Finally, sustainable development in the construction industry is to meet people's demand for a better living environment, to reduce building energy consumption to the maximum extent, and to save costs.

In order to promote sustainable development in the construction sector, there are three concepts of green building, green construction, and sustainable construction that have been studied in academics and practised in industry. Green buildings are defined as high-environmental performance buildings that provide comfortable, healthy, eco-friendly, and economical life and workplaces. The performance requirements of green buildings focus on the final products of buildings in operation and service stages, sometimes considering the whole building life cycle. The performance measurements of green buildings are generally from energy and water consumption. Moreover, the Green Construction Guideline issued by China's Ministry of Construction defined:

> On the premise of ensuring quality, safety and other basic requirements, scientific management and technological progress should be used in engineering construction, to maximise the conservation of resources and reduce the construction activities which will bring negative impacts on the environmental, and to achieve the goal of four savings (energy, land, water and materials) and environmental protection.
>
> (Shi et al., 2013)

Therefore, green construction focuses on environmental construction processes and sites, where the consumption of energy, land, water, and materials of construction projects is mainly assessed. Finally, Gehlot and Shrivastava (2022) employed the concept of sustainable construction which is defined by the UN Department of Economic and Social Affairs as "The construction that brings about the required performance with the least unfavourable ecological impacts while encouraging economic, social and cultural improvement at local, regional and global level". Accordingly, sustainable construction focuses on the process from economic, environmental, and social viewpoints particularly in the project construction stages, sometimes considering the whole construction industry. Compared to green construction, sustainable construction also evaluates social responsibility and win-win economic performance. Social responsibilities are long-term commitments where construction projects, corporations, and industry contribute to developing the sustainability of the society in which they operate by behaving ethically. The win-win economic performance displays that the economic success of construction will produce benefits for developers, designers, builders, suppliers, services, workers, and other project stakeholders.

In this book, sustainable construction will be redefined and studied. Sustainable construction refers to the construction process and surroundings from environmental and social viewpoints. It aims to minimize the environmental impact of construction activities, promote resource and energy efficiency, reduce waste and carbon footprints, and create a healthy and safe work environment, ensuring the improvement of the environmental and social sustainability performance of project construction, along with the achievements of the economic sustainability performance. Sustainable construction focuses on the practices during the construction process, although considering the life cycle impact of projects on promoting sustainable construction performance. The performance measurement indicators for sustainable construction include the performance indicators of green construction and the social responsibility of project construction. The differences among green building, green construction, and sustainable construction are demonstrated in Table 1.2.

Table 1.2 Differences among green building, green construction, and sustainable construction

Aspects	*Green building*	*Green construction*	*Sustainable construction*
Focus on	Product. Focus on the final construction results—buildings.	Process and sites. Focus on environmental construction processes and sites.	Process and surroundings. Focus on the construction process and surroundings from environmental and social viewpoints.
Stages	Mainly in the operation and service stage, sometimes considering the life cycle of buildings	Construction stage	Mainly in the construction stage for projects, particularly the life cycle of projects
Primary performance measurement indicators	Energy and water consumption of buildings	Waste, energy, water, and materials of construction projects	Including green construction indicators, construction surroundings, and social responsibility performance

The significance of implementing sustainable construction embodies reducing energy and water consumption; decreasing emissions and pollution; improving waste recycling; using low-carbon, recyclable, and renewable materials; and minimizing construction activities' impacts on ecosystems and wildlife. For managing waste, sustainable construction practices contribute to waste disposal actions such as avoiding, reducing, reusing, and recycling. Moreover, sustainable construction can benefit to find solutions to diminish the consumption of energy resources, materials, and land during the construction phase of projects. More importantly, good sustainable construction practices can improve construction cost efficiency and productivity and benefit environmental, social, and economic performance. Therefore, promoting sustainable construction practices can support environment recovery, improve material usage efficiency, enhance construction WHS performance, drive environmental sustainability, improve construction productivity, and benefit climate change mitigation objectives.

1.5 The scope of this book

The Asian construction industries are selected as a case study, as the Asia region accounts for the largest share of the global construction industry and will continue to be the main development engine of the world. Another important reason is that Asia has the largest developing countries in the world such as China and India, and sustainable construction for developing countries is more important than developed countries as developing countries generally require large-scale construction in the built environment and its physical infrastructure.

The aim of this book is mainly to improve sustainable construction performance, particularly for environmental and social sustainability, focusing on sustainable construction management, methods, technologies, techniques, and materials. Empirically underpinned by experiences, observations, and research from the field, drawing upon an analysis of sustainable development solutions from the Asian construction industries, this book presents primary sustainable construction systems of how to promote sustainable development in the construction industry during a life cycle management process and through key practical drivers. Moreover, the book aligns several contemporary discourses—closed-loop production, green building, green economy, industrial ecology, industry social responsibility, life cycle analysis and management, and sustainable construction management—into establishing an interdisciplinary dialogue both to enhance scholarship and to inform practice. Accordingly, the sustainable development within the construction industry discussed in this book encompasses the following chapters:

Chapter 1: Sustainable development in the construction industry: Introduction

This initial chapter establishes the context and conceptual scenes for this book. The reasons for and implications of promoting sustainable development in construction industries are emphasized. It advocates for a sustainable development manifesto in response to the challenges between sustainability and development in the Asian context. Particularly, the definition and scope of sustainable construction are developed for this book.

Chapter 2: Sustainable construction in Asia: Overview

This chapter provides an overview of the current status, pressures, and requirements for sustainable development in the Asian construction sector. It critically analyses sustainable construction practices across various Asian countries, examining governmental and industry initiatives to enhance sustainability. The chapter aims to offer a solution framework for decision-makers to support sustainable development in construction through political advocacy, regulatory frameworks, economic incentives, education, technical innovation, and social engagement.

Chapter 3: Construction project management methods towards sustainable development

This chapter introduces advanced project management methods for sustainable development and highlights best practices in applying these methods within the Asian construction industry. Topics include life-cycle management, sustainable procurement, risk management, lean construction, and integrated project delivery methods, with detailed discussions on management cases from a project management perspective for sustainable construction.

Chapter 4: Smart construction technologies towards sustainable development

Technological innovations and applications play an important role for achieving sustainable development, as they can encourage environmental issues without discouraging production growth and social progress. The chapter provides smart construction technologies that have the ability to improve construction sustainability, including 3D printing construction, BIM-based sustainable construction, industrialized construction, prefabricated construction, and digital construction technologies such as eXtended reality, machine learning, digital twins, big data, blockchain, Internet of Things, and cloud construction technologies.

Chapter 5: On-site construction techniques towards sustainable development

Focused on on-site construction techniques, this chapter outlines a framework for sustainable construction management, organizational structures, plans, resource management, environmental practices, and social sustainability commitments. It details specific measures and actions within the on-site construction process, showcasing sustainable practices in the Asian construction sectors.

Chapter 6: Introduction to sustainable construction materials

Sustainable construction is closely linked to the materials used in building projects. This chapter explores a broad array of primary sustainable construction and green building materials, with a particular focus on recycled and reused materials, as well as waste materials in concrete construction. Additionally, it offers recommendations on how these materials can be effectively utilized to promote building sustainability performance within the construction industry.

In summary, sustainable development in the construction industry emphasizes managing all aspects of the sector to ensure the long-term success of projects, companies, and industries. It aims to provide people with healthy, suitable, and efficiently used spaces that benefit society, the environment, the economy, governments, and communities. This book seeks to bridge the gap between academia and the profession by exploring new sustainable development opportunities in the construction industry, ultimately preparing a sustainable world for future generations.

References

Bon, R. (1992). The future of international construction: Secular patterns of growth and decline. *Habitat International, 16*(3), 1119–1128.

Buckley, M., Zendel, A., Biggar, J., Frederiksen, L., & Wells, J. (2016). *Migrant work & employment in the construction sector*. International Labour Organization.

Gehlot, M., & Shrivastava, S. (2022). Sustainable construction Practices: A perspective view of Indian construction industry professionals. *Materials Today: Proceedings, 61*, 315–319.

GlobalABC/IEA/UNEP (Global Alliance for Buildings and Construction, International Energy Agency, and the United Nations Environment Programme). (2020). *GlobalABC regional roadmap for buildings and construction in Asia: Towards a zero-emission, efficient and resilient buildings and construction sector.* IEA.

Huld, A. (2022). *China infrastructure investment in 2022—Can it stimulate economic growth?* Published by China Briefing in www.china-briefing.com/news/china-infratsructure-investment-in-2022-spurring-economic-growth/.

IEA. (2019). *World energy outlook 2019.* OECD Publishing. https://doi.org/10.1787/caf32f3b-en.

Shi, Q., Zuo, J., Huang, R., Huang, J., & Pullen, S. (2013). Identifying the critical factors for green construction—an empirical study in China. *Habitat International, 40*, 1–8.

United Nations. (1987). *Report of the world commission on environment and development: Our common future.* https://sustainabledevelopment.un.org/content/documents/5987our-common-future.pdf

United Nations. (2013). *World economic and social survey 2013 sustainable development challenges.* https://sustainabledevelopment.un.org/content/documents/2843WESS2013.pdf

United Nations. (2015). *Sustainable development goals.* www.un.org/sustainabledevelopment/sustainable-development-goals/

United Nations Environment Programme. (2021). *2021 Global status report for buildings and construction: Towards a zero-emission, efficient and resilient buildings and construction sector.* Nairobi. https://globalabc.org/sites/default/files/2021-10/GABC_Buildings-GSR-2021_BOOK.pdf

2 Sustainable construction in Asia

Overview

2.1 Introduction

Asia is the largest continent in the world, which occupies about 30% of the Earth's total land area. It also has the world's largest population—some 4.7 billion people in 2022—approximately 60% of the total population in the whole world. Particularly, the population is still growing, such that the growth rate is 0.83% per year in 2020. The largest continent could provide the land area for building construction. The highest population and growth birth rate not only display the biggest demand for building areas but also provide construction workers for the construction market. Accordingly, these conditions could provide potentials and support for the development and prosperity of the building construction industry in Asia.

In Asia, most countries can be characterized as the developing countries. For example, China, India, and Indonesia are the world's most economically developing countries with higher economic growth rates. For instance, the International Monetary Fund in July 2021 predicted that Asia would grow at 7.5% and 6.4% in 2021 and 2022, respectively, compared with 6.0% and 4.9% for the world. Economic growth and social development in developing countries are especially vital, which results in the increase of energy consumption and carbon emissions. For example, according to the "Statistical Review of World Energy 2021", Asia-Pacific is home to some of the world's largest carbon emitters, which accounted for 52% of global carbon dioxide emissions in 2020. As a consequence, how to sustainably develop the economy and society has been becoming a serious challenge for most of the Asian countries. Particularly, the COVID-19 pandemic was a stumbling block for sustainable development in the whole world.

In the report *Green Growth Index 2020*, from 2005 to 2020, the green growth in Asian countries displayed lower performance than countries of Europe, Oceania, and Americas, and is only higher than that in African countries. The Green Growth Index is measured by the four dimensions of efficient and sustainable resource use, natural capital protection, green economic opportunities, and social inclusion, aiming to achieve sustainability targets including the Sustainable Development Goals (SDGs), the Paris

DOI: 10.4324/9781003202660-2

Climate Agreement, and the Aichi Biodiversity Targets. The different countries evidently performed various green growth scores in Asia. For example, the green growth in Eastern Asia is more than 10%, 20%, 30%, and 60% of South-eastern Asia, Western Asia, Southern Asia, and Central Asia, respectively. The reason is because the green economic opportunities are highly developed due to green investment and green employment in Eastern Asia. For example, in Asia, the best three countries are Japan, Georgia, and China, where green investment and income levels are the main contributing factors towards higher performance in green economic opportunities. Moreover, compared to African nations, the Asian developing countries are weaker in the dimension of natural capital protection, as a result of the priority of industrialization over conservation.

The Sustainable Development Report 2021 presents the performance of countries to achieve the SDGs in the whole world. Japan and Korea Rep. performed the best rankings of 18 and 28 out of 165 countries. Thailand is ranked 43, which is the best developing country in Asia. Although East Asia and Southern Asia have improved their performance towards the SDGs than any other region since 2020, countries in the region have distinct sizes, populations and economic development levels, which are facing a corresponding range of challenges in achieving the goals. For example, many countries need a significant acceleration of progress in promoting the goal of Life on Land by protecting biodiversity and threatened species. Compared to the OECD's (Organization for Economic Co-operation and Development) member countries, there is a significant amount of effort and long-term commitment to promoting sustainable development in Asian countries. The Asia and the Pacific SDG Progress Report 2020 pointed out that "urgent action is needed to protect the environment, reduce the risk of natural disasters, and take climate action. Therefore, this book will focus on how to achieve the SDGs in Asia.

2.2 Case studies of sustainable construction in Asia: a review

2.2.1 Central Asia

Kazakhstan

Kazakhstan's construction sector is navigating the complex transition towards sustainable construction, which is influenced by internal and external drivers and is faced with significant barriers. Workshop among construction professionals identified environmental issues and the efficient use of construction materials and resources as critical drivers for adopting sustainable construction practices. This reflects a growing awareness and willingness among state bodies and stakeholders to embrace sustainability measures, driven by the urgent need to address environmental challenges and unsustainable construction patterns.

Despite this momentum, economic aspects, followed by governmental awareness and education-related barriers, significantly hinder the adoption. Tokbolat et al. (2019) highlighted the vital role of addressing environmental and social aspects as critical drivers for sustainable construction in Kazakhstan. It provides insights into the barriers and drivers specific to developing countries, particularly those in Central Asia with similar socioeconomic contexts. It underscores the necessity for targeted policy recommendations and strategic interventions to overcome these barriers, suggesting that a context-oriented approach is crucial for advancing sustainable construction practices within Kazakhstan's construction industry.

A range of barriers hinders the journey towards sustainable construction in Kazakhstan. A significant impediment is the need for more active promotion and support from the government, including the absence of robust regulations, policies, and incentives specifically geared towards sustainable construction. Economic challenges also play a crucial role, with the high cost of sustainable options and more extended payback periods deterring stakeholders. The priority often given to immediate economic needs overshadows the long-term benefits of sustainability. Moreover, there needs to be more understanding and awareness among stakeholders about the benefits and technologies associated with sustainable construction. The construction sector needs more professional capabilities and competence in sustainability, further exacerbated by a lack of adequate education and training for construction professionals. Additionally, these challenges are compounded by the limited availability of green suppliers and information, the absence of demonstration projects, and a reluctance to shift from conventional methods. These barriers collectively impede the widespread adoption of sustainable practices in Kazakhstan's construction sector.

A comprehensive and multifaceted approach is essential to enhance sustainable construction in Kazakhstan. Key solutions include robust government-led promotion and support for sustainable construction, emphasizing the development and strict enforcement of relevant regulations and policies. Providing government incentives and financial support can address the economic challenges associated with sustainable construction. Education and training programs for construction professionals are crucial for building competence and understanding of sustainable practices. Raising awareness about the benefits of sustainable construction and promoting sustainable technologies will help shift the industry's perception and approach. Encouraging the use of green suppliers and sustainable materials, along with implementing demonstration projects, can showcase the practical benefits of sustainable construction. Prioritizing sustainability in bidding and procurement processes, engaging clients and stakeholders in sustainable practices, and transitioning from conventional to sustainable construction methods are other essential steps. These measures, when effectively implemented, can significantly contribute to overcoming the current barriers and propel Kazakhstan towards a more sustainable and environmentally responsible construction sector.

2.2.2 Eastern Asia

China

In China, the push towards green construction is critical to the country's broader environmental goals and SDGs. Wang et al. (2021) identified significant barriers to adopting green construction practices in their comprehensive study that surveyed industry stakeholders. The research highlights various challenges, including management barriers, technological hurdles, economic factors, and awareness issues as being at the forefront. The cost implications associated with green construction, alongside unfamiliarity with green technologies, emerge as the principal obstacles impeding broader implementation within the industry. Despite these challenges, there is an evident commitment to overcoming them, reflected in the initiation of various programs aimed at bolstering the capacity of the construction sector to adopt sustainable practices. This includes training and certification programs for green construction professionals and efforts to establish clearer green building standards and guidelines. However, the findings underline the necessity for a more integrated approach that encompasses technological and managerial adjustments, significant policy interventions, and educational campaigns to enhance understanding and support for green construction practices. The study's insights into the Chinese context underscore the complex interplay of factors that must be addressed to promote sustainable construction, suggesting a pathway that involves strengthening government policies, industry practices, and public awareness to achieve sustainable construction objectives.

In China, sustainable construction is challenged by various barriers. The most prominent include the need for more government incentives, high additional costs associated with green construction technologies (GCTs), and dependence on traditional construction methods. Other significant barriers include a shortage of technological training for project staff, conflicts of interest among stakeholders, imperfect green construction technological codes and standards, and a lack of quantitative assessment tools for green performance. Further issues include low compatibility with other construction technologies, limited availability of green materials and equipment, project delays caused by sustainable construction implementation, and increased risk and uncertainties with GCT adoption. The lack of attention from senior managers and insufficient environmental awareness among organization managers and technicians exacerbate these challenges, alongside low labour quality.

China needs to adopt a multifaceted approach to overcome the barriers to sustainable construction. Enhancing government incentives and policy support for GCT adoption is essential. Addressing the high costs through financial and non-financial incentives and reducing dependence on traditional construction technology by showcasing the advantages of GCTs can facilitate change. Providing extensive technological training for project staff, resolving conflicts of interest among stakeholders, developing green construction codes and standards, and creating assessment tools for green performance are critical steps. It is also necessary to improve the compatibility of GCTs with other construction technologies, ensure the availability of green materials and equipment, and mitigate project delays caused by sustainable construction practices. Moreover, reducing incremental risks associated with sustainable construction, encouraging senior managers to focus on GCTs, raising environmental awareness, and improving labour quality through education and training initiatives are expected to advance sustainable construction practices in China. These combined efforts could transform the construction industry, making it more sustainable and environmentally friendly.

2.2.3 South-eastern Asia

Indonesia

In Indonesia, sustainable construction faces various barriers despite the country's growing awareness and initiation towards green building practices. The study conducted on implementing sustainable construction in Palembang, Indonesia, underscores the challenges spanning economic, technical, regulatory, and socio-cultural aspects (Susanti et al., 2019). Among the highlighted obstacles are the limited number of trained or certified workers and the need for more effective communication between project stakeholders. These findings suggest a significant gap in skilled labour and project management practices that align with sustainability goals. Despite these hurdles, Indonesia has made strides in incorporating environmental considerations into construction projects, as seen in seminars and training on green building concepts. However, the implementation predominantly touches on environmental aspects, overlooking sustainability's comprehensive social and economic dimensions. This partial adoption indicates a broader issue prevalent in developing countries, where economic and social factors often overshadow environmental sustainability. The need for a more integrated approach to sustainable construction in Indonesia is evident, calling for enhanced regulatory support, financial mechanisms, and educational efforts to overcome the identified barriers and promote a holistic adoption of sustainability in the construction sector.

The main barriers to sustainable construction in Indonesia are multifaceted. The workforce's limited skills and expertise significantly impede the adoption of sustainable practices. Ineffective communication among project teams is another major hurdle affecting the coordination and efficient execution of sustainable construction projects. Economic challenges, including potential increases in total project costs and duration, pose significant obstacles. Indonesia's limited economic capacity, insufficient regulations and standards promoting sustainable construction, and low implementation of regulations for skills and expertise training hinder progress. Additionally, the need for more preparedness by companies for sustainable construction demands, limited availability of suitable equipment and materials, scarce guidelines or modules on sustainable construction, and a general preference for conventional methods over sustainable ones are notable issues. The limited demand for projects using sustainable construction, coupled with a need for more awareness about the advantages of sustainable methods and misconceptions that equate sustainable construction solely with green construction, adds to these challenges.

Several solutions can be implemented to overcome these barriers and advance sustainable construction in Indonesia. Improving training and certification programmes for workers in sustainable construction practices is crucial. Enhancing communication and coordination among project teams can streamline the implementation of sustainable construction methods. Addressing financial and economic challenges through efficient project management and budgeting can make sustainable construction more viable. Developing and enforcing supportive regulations and standards for sustainable construction is essential for guiding industry practices. Increasing company preparedness and adaptability to sustainable construction demands, expanding the availability of sustainable construction materials and equipment, and creating comprehensive guidelines and educational resources on sustainable construction will support industry transition. It is vital to promote sustainable construction methods over conventional ones through awareness campaigns and encouraging demand for sustainable construction projects among stakeholders. Educating the construction industry and the public on the benefits and feasibility of sustainable construction can change perceptions and drive adoption. Differentiating sustainable construction approaches to highlight its broader scope and mitigating perceived risks through evidence-based strategies and case studies will further reinforce the move towards sustainable construction practices in Indonesia.

Malaysia

In Malaysia, the transition towards sustainable construction, mainly green building, faces notable barriers, including low market demand, a lack of awareness among stakeholders, insufficient green building regulations and codes, and a need for more financial incentives and training on green technologies (Wong et al., 2021). Despite these challenges, initiatives like the Green Building Index (GBI) have been developed to encourage environmentally friendly construction adapted to Malaysia's unique climate and needs. However, the adoption rate remains modest, highlighting the critical need for increased education and awareness about the benefits of green building, alongside more robust government support through policy development, incentives, and targeted training programs. This comprehensive approach is vital for overcoming obstacles and accelerating Malaysia's journey towards sustainable construction practices.

The primary barriers to the adoption of sustainable construction in Malaysia are multifaceted. The most significant challenge is the low market demand for green buildings and the need for green building regulations and codes.

There is also a noticeable need for more subsidies for green technology and insufficient training in green building technologies. A general lack of awareness of green building technologies' existing incentives and benefits further complicates matters. Environmental awareness among developers and consultants needs to be improved, and more adequate information and databases on green technologies should be available (Saharuddin et al., 2022). Resistance to change by supply chain agents, high initial and maintenance costs, limited finance support for upfront costs, clients' preference for instant payback, and risks associated with new green technologies are other notable barriers. These challenges are compounded by the need for more understanding and awareness of sustainable construction concepts, making it difficult for the industry to transition to more sustainable practices.

Several solutions can be implemented to address these challenges and enhance sustainable construction in Malaysia. Increasing market demand through client education and awareness programs can stimulate interest in green buildings. Implementing and enforcing green building regulations and codes, providing subsidies and financial incentives for green technology, and offering training programs focused on green building technologies are crucial. Promoting the long-term benefits and incentives of green building, developing comprehensive information databases on green technologies, and encouraging stakeholder collaboration can help overcome resistance to change. Addressing the high-cost barrier through financial support and cost-effective green solutions is essential. Additionally, developing standardized databases for green building information storage and conducting further research on local context and client preferences regarding green building can provide valuable insights for tailoring sustainable construction practices to Malaysia's unique needs. These initiatives, coupled with efforts to reduce the time and cost investment required for sustainable construction, provide education and training to contractors, and introduce government incentives, and can significantly advance the green movement in Malaysia's construction industry.

Vietnam

Vietnam's journey towards sustainable construction is marked by significant challenges, including managerial competence, limited availability of sustainable materials and technologies, resistance to change, and a need for government incentives. These barriers highlight a complex environment where educational gaps, technological scarcities, and inadequate regulatory frameworks slow down the adoption of sustainable practices. To navigate these obstacles, Vietnam requires targeted efforts to enhance project management skills, develop a local market for sustainable materials, foster a culture of sustainability, and strengthen governmental support through incentives and policies. This multifaceted approach emphasizes the need for comprehensive strategies encompassing education, technology, cultural change, and policy reform to advance sustainable construction practices within Vietnam, reflecting broader challenges and solutions in the developing world's pursuit of sustainability in the construction sector.

Pham et al. (2019) conducted a study involving firm directors and project managers through literature review and surveys, which highlighted critical barriers at both the firm and project management levels. The main barriers to sustainable construction in Vietnam are multifaceted. Incompetence of project managers is a significant challenge that hinders the effective implementation of sustainable practices. The limited availability of sustainable materials and technologies also poses a substantial obstacle, restricting the scope for sustainable construction. A cultural resistance to change, with a tendency to maintain current practices, further impedes the transition towards sustainability. The lack of government incentives for sustainable construction projects and the overall low level of implementation of sustainable practices within the industry compound these issues. These barriers reflect the need for a holistic approach to overcome the challenges and encourage sustainable practices in Vietnam's construction sector.

Several solutions are proposed to address these challenges. The first crucial solution is to improve the competence of project managers through regular training, deep inter-professional collaboration, and multidisciplinary practices in sustainable construction. This includes building knowledge and leadership competencies specifically tailored to sustainable construction. Addressing the limited availability of sustainable materials and technologies is essential, as it encourages the development of new, energy-efficient, and environmentally friendly construction materials. Adopting off-site construction methods like prefabrication can offer additional benefits. Cultivating an innovative culture within construction firms requires leaders and managers to promote new sustainable practices actively. Encouraging a shift in traditional mindsets and habits towards sustainability is necessary to foster a more sustainable construction industry in Vietnam. If effectively implemented, these solutions could lead to significant advancements in sustainable construction practices, benefiting the country socio-economically and environmentally.

2.2.4 Southern Asia

India

India's construction sector is actively navigating towards sustainable practices, underpinned by the government's push towards green building initiatives and integrating international sustainability frameworks. Despite facing challenges such as misconceptions about the cost implications of green building and varying adoption rates, efforts such as developing green building codes, promoting energy efficiency, and waste minimization practices signify a robust move towards environmental stewardship. However, the path forward requires addressing significant barriers, including low market demand, insufficient regulatory frameworks, and the need for enhanced stakeholder awareness and technical expertise. A comprehensive strategy encompassing regulatory reforms, financial incentives, and educational initiatives is essential for India to embrace sustainable construction practices fully, aligning economic growth with environmental sustainability in its quest to build a greener future.

Several barriers hinder the transition to sustainable construction in India. Significant challenges include low general literacy rates, particularly regarding environmental issues, and a cultural deterioration that overlooks traditional respect for the environment (Arif et al., 2009). Poverty often limits the ability to cover the upfront costs of implementing green principles. The negative influence of technology on building construction, coupled with a lack of optimal green construction education in schools and colleges, further impedes progress. Additional challenges include slow technology transfer from countries with advanced green practices, over-construction in construction leading to increased waste, and resistance to moving away from traditional practices to sustainable methods. The construction sector's contribution to environmental degradation underscores the urgency of addressing these barriers to promote sustainable construction practices.

To enhance sustainable construction in India, a multifaceted approach is necessary. This includes enhancing environmental literacy and awareness, promoting a cultural shift towards environmental responsibility, and providing financial support to offset the costs of green construction. Encouraging the adoption of environmentally friendly technologies in building construction and integrating green education into academic curriculums are crucial steps. Accelerating the transfer and adoption of advanced green technologies can help bridge the technological gap. Advocating for optimal sustainable practices to minimize waste in construction is essential. Facilitating the transition from traditional to sustainable construction practices is vital for the sector's long-term sustainability. Additionally, enhancing management commitment to adopting green practices, addressing perceptions of higher costs, improving knowledge of green products, strengthening collaborations, and encouraging market demand for green products through policy incentives and public awareness campaigns are vital strategies to overcome the barriers in sustainable construction. These combined efforts can significantly advance India's journey towards environmentally responsible and economically viable construction practices.

Iran

In Iran, sustainable construction and project management have emerged as critical areas of focus within the construction industry, acknowledging the sector's substantial impacts on environmental sustainability and the need for systemic change. Fathalizadeh et al. (2021) discussed the barriers that hinder the adoption of sustainable project management practices in the Iranian construction sector, specifically, project context, knowledge, investment, community, and strategy. These challenges collectively represent the multifaceted challenges to implementing sustainable practices in construction project management. The barriers to sustainable construction in Iran are complex and multifaceted, ranging from an unstable economy that affects the availability of sustainable resources to a need for greater political will and commitment to sustainability. There is a need for a more precise vision and planning when it comes to integrating sustainability, and a lack of coordination between industry sectors, academia, and governmental agencies exacerbates the issue. A general lack of awareness and understanding regarding sustainability further impedes progress. These barriers are compounded by economic and regulatory-dependent challenges, including inadequate cooperation among various stakeholders and the absence of systematic approaches to achieve sustainability goals. These obstacles suggest that achieving sustainable construction in Iran requires changes at both the project and industry levels, as well as a broader transformation in the socio-economic and regulatory landscape.

A series of strategic interventions is required to enhance sustainable construction in Iran. Stabilizing the economy is crucial to supporting sustainable construction initiatives and increasing political will and commitment towards sustainable practices. Developing a clear vision and comprehensive plans for sustainability integration is essential for guiding industry efforts. Enhancing coordination and cooperation among industry sectors, academia, and government can foster a more collaborative and practical approach. Raising awareness and understanding of the benefits and practices of sustainability is vital to changing perceptions and attitudes. Addressing economic and regulatory barriers through reforms and incentives is necessary to create a conducive environment for sustainable construction. Encouraging collaborative efforts among stakeholders can lead to more sustainable advancements, while implementing systematic and holistic approaches to sustainability in construction will ensure that efforts are comprehensive and aligned with broader national goals. When effectively implemented, these solutions can contribute to advancing sustainable construction in Iran, thereby supporting the country's overall economic development and environmental sustainability.

Pakistan

Pakistan's construction sector faces a substantial challenge in adopting sustainable construction practices, particularly energy management practices (EMPs), despite their significant potential for cost efficiency and environmental protection. Iqbal et al. (2021) discussed sustainable construction through EMPs within the Pakistani construction industry, where the identified critical barriers consist of the lack of communication and collaboration among project stakeholders; attitudinal, cultural, and behavioural resistance to EMPs; lack of top management support, and absence of financing schemes for energy management technology.

Pakistan's sustainable construction landscape is impeded by several barriers. There is a significant gap in communication and collaboration among stakeholders in construction projects, which hampers the effective implementation of EMPs. Additionally, there is cultural and attitudinal resistance towards adopting these practices, often stemming from traditional building methods and practices. The lack of top management support and interest in energy management issues further exacerbates this problem. Moreover, the construction sites often lack the conditions to implement EMPs effectively. Financial constraints are another major hurdle, with limited financing schemes available for implementing energy management technology and the higher costs associated with adopting EMPs. Government support and incentives to promote these practices are also lacking. The dearth of innovative energy management technology and customer demand for sustainable construction further complicates the scenario. Lastly, government policies and legislation require significant improvement to support and promote sustainable construction practices effectively.

Several measures need to be adopted to enhance sustainable construction in Pakistan. Improving communication and collaboration among stakeholders is crucial to effectively implement sustainable practices. Cultivating a positive attitude towards sustainability through awareness and education can help overcome cultural and attitudinal barriers. Encouraging top management to support and invest in energy management can have a transformative impact. Creating conducive conditions and policies for implementing EMPs in construction projects is essential. Developing financing schemes and incentives for adopting energy-efficient technologies can address the financial barriers. Government support in the form of incentives and subsidies can significantly boost the adoption of sustainable practices. Introducing and promoting innovative energy management technologies will also play a vital role. Enhancing customer awareness and demand for sustainable construction can create a market-driven push for greener building practices. Finally, implementing effective government policies and legislation will provide the necessary framework and support for sustainable construction to thrive. Therefore, addressing these factors can pave the way for a more sustainable construction sector in Pakistan, balancing economic growth with environmental responsibility.

2.2.5 Western Asia

Kuwait

In Kuwait, the journey towards sustainable construction is embryonic, grappling with significant challenges such as a pervasive lack of awareness among industry stakeholders and insufficient government initiatives to foster green building practices. AlSanad (2015) emphasized the findings from a study on sustainable construction in Kuwait, highlighting the critical need for robust government support through comprehensive policies, standards, and incentives to catalyse the adoption of sustainable practices within the construction sector. Although there is an acknowledgement of the environmental, economic, and social benefits that sustainable construction can bring, the transition is hindered by systemic barriers, including the heavy subsidization of electricity, which diminishes the financial incentives for energy-saving measures. To move forward, Kuwait requires a concerted effort involving enhanced stakeholder education, governmental incentives, and policy reforms to integrate sustainable construction practices into the mainstream, thereby aligning with the global environmental sustainability effort.

The adoption of sustainable construction methods in Kuwait faces several barriers. A significant impediment is the need for more awareness about the benefits and practices of green building. This is compounded by insufficient government support and incentives, which are crucial for the growth of sustainable construction. Economic conditions, perceived risks associated with implementing new practices, and the prevailing notion that green building is expensive deter stakeholders. Additionally, the lack of definitive evidence regarding the benefits of green construction further exacerbates this reluctance, hindering widespread adoption of such methods. Furthermore, a limited number of developers are currently undertaking green building projects, which limits the visibility and perceived feasibility of such practices. Traditional construction practices are deeply rooted, and there is a general unwillingness to shift from these conventional methods. Finally, the need for qualified staff for green construction and the absence of specific rules and legislation in Kuwait hinder the progress towards sustainable construction.

To enhance sustainable construction in Kuwait, a multifaceted approach is essential. Critical solutions include the implementation of educational programs aimed at raising awareness about green construction and its benefits. It is crucial to establish and enforce rules and legislation that promote sustainable construction practices. The development of sustainable construction guidelines and standards can provide a clear framework for industry practitioners. Providing economic incentives can stimulate interest and investment in green practices. Conducting research and publicizing the tangible benefits of green construction will help counteract scepticism and uncertainty. Encouraging the private sector to participate in green building projects can diversify and expand the adoption of sustainable practices. Promoting a change in mindset towards sustainable practices is necessary for long-term transformation. Increasing the availability of qualified professionals in sustainable construction through education and training will address the current skills gap. Developing and implementing governmental policies and incentives targeted explicitly at green construction can provide regulatory and financial support. Lastly, encouraging collaboration between the public and private sectors to support green initiatives can create a more cohesive and practical approach to sustainable construction in Kuwait.

Oman

Oman's construction sector is actively evolving towards integrating sustainable practices amidst its rapid infrastructure development, yet faces significant challenges in adopting green construction widely. The Sultanate recognizes the importance of sustainability for future generations and has initiated various projects and policies to encourage green building practices. Despite this, the implementation of sustainable construction faces several hurdles, primarily due to a lack of demand for green construction and insufficient governmental pressure. Oman's efforts include substantial investments in infrastructure and utilities, aiming to balance intensive development with environmental conservation (Saleh & Alalouch, 2015). The country's environmental law exists but needs more clarity in effectiveness, signalling a need for more decisive actions and comprehensive policies. Initiatives such as the Oman Eco House Design Competition and the Oman Green Building Council establishment in 2012 demonstrate a move towards promoting green construction concepts. However, these efforts are still in their infancy, necessitating greater governmental involvement, enhanced professional body activity, and a shift in construction responsibilities from engineers to architects to truly embed sustainable practices in Oman's construction industry. The country's approach to sustainable construction indicates a broader regional trend in the Gulf Cooperation Council countries, where rapid construction growth presents opportunities and challenges for integrating green elements into projects.

The path to sustainable construction in Oman is riddled with several obstacles. A notable constraint is the low demand for green construction, stemming from a need for more awareness and understanding of its principles among industry professionals and the broader public. Economic factors also play a crucial role; the higher perceived costs of green construction methods act as a deterrent to stakeholders who are accustomed to traditional construction practices. Moreover, the traditional construction methods are deeply ingrained in Omani culture, further complicating the transition to more sustainable practices. Furthermore, the need for more vigorous government enforcement and comprehensive policies geared towards sustainable construction is evident. The current regulatory framework needs to incentivize or mandate green practices sufficiently. Lastly, there needs to be more emphasis on the architectural aspect of construction, particularly in integrating sustainable construction principles from the onset of projects.

To enhance sustainable construction in Oman, a multifaceted approach is necessary. Foremost, increasing government participation is crucial, which includes developing and strictly enforcing local green construction policies and standards to catalyse a shift in industry practices. Educating practitioners and stakeholders about the long-term benefits and practicalities of sustainable construction is another vital step. This education can help mitigate economic concerns by highlighting the long-term cost-effectiveness and environmental benefits of green building practices. Breaking away from traditional construction methods and encouraging innovative, sustainable alternatives can foster a more environmentally conscious construction culture. Additionally, empowering architects to take a more significant role in the construction process can ensure that sustainable practices are integrated from the project's inception. Addressing these issues, Oman can pave the way for a more sustainable and environmentally responsible construction sector, aligning with global trends and contributing to the country's SDGs.

Qatar

Sustainable construction in Qatar is at a crucial juncture, grappling with challenges despite the global advancement in green construction technology. A comprehensive study involving 129 building experts highlighted the significant environmental impact of Qatar's construction industry, notably its substantial energy consumption, resource use, and contribution to CO_2 emissions (Ahmed et al., 2023). These findings underscore the urgent need to transition to sustainable construction practices as a mitigation strategy against these adverse impacts. However, the path to sustainable construction in Qatar is fraught with obstacles. The primary barriers identified include inadequate green construction codes, a need for more information and understanding about new sustainable techniques, and difficulties in acquiring the necessary skills and expertise. Furthermore, the market's limited investment in green projects due to low demand represents a significant challenge. These issues, while particularly pronounced in Qatar, resonate across the Gulf region and beyond, indicating a broader regional trend in the sustainable construction landscape.

According to the research conducted by Ahmed et al. (2023), the top five risks associated with sustainable construction in Qatar are as follows: (1) building codes and regulations lagging behind the latest technologies in green and sustainable practices; (2) insufficient awareness and training; (3) a lack of educational curricula focused on sustainability and green building; (4) difficulty in finding skilled expertise in green technologies; and (5) low investment in green initiatives due to limited demand within the local construction industry. Additionally, responsibility for addressing these critical risks should be distributed among key stakeholders, as indicated by expert interviews. The ranking of these stakeholders to share the responsibility is as follows: government bodies, consultants, contractors, suppliers, and other relevant parties.

Implementing targeted solutions can overcome the barriers to sustainable construction in Qatar. Developing and enforcing comprehensive green construction codes and regulations can provide a robust legal framework for industry practices. Enhancing information dissemination and understanding new sustainable techniques are crucial for industry-wide adoption. Facilitating skill development and expertise in green construction will address the current gaps in professional capabilities. Encouraging investment in green projects through market stimulation strategies can overcome the challenge of limited financial backing. Addressing the high-cost issues, including stabilizing currency fluctuations, can make sustainable construction more financially viable. Improving cost-effectiveness and efficiency in project timelines is essential for demonstrating the practical benefits of sustainable practices. Increasing access to and utilization of renewable resources can significantly reduce the environmental footprint of construction projects. Lastly, implementing educational programs and training in sustainable construction practices will ensure a well-informed and capable workforce, which is crucial for successfully implementing sustainable construction projects in Qatar.

2.3 Barriers to sustainable construction in Asia

Table 2.1 summarizes and displays the barriers encountered by various Asian countries in sustainable construction practices, delineating the myriad challenges inhibiting the adoption of sustainable construction across the Asian continent. The table categorizes these challenges into economic barriers such as high costs and lack of financing; social and cultural barriers including resistance to change and traditional practices; political and regulatory barriers highlighted by insufficient government enforcement and the need for policy improvement; and technical and knowledge barriers such as the lack of innovative technology and limited knowledge. Overall, high costs are the most common barriers in practice. Moreover, the frequencies of these barriers are displayed in the last column of Table 2.1. Some obstacles are rooted in the unique economic, cultural, and political contexts of each region, while others are more common across different areas. This categorization highlights the need for holistic strategies that address the specific needs and constraints of various Asian countries in promoting sustainable construction practices.

In Western Asia, represented by countries like Oman, Kuwait, and Qatar, a prominent trend is the need for enhanced government support and comprehensive policies in sustainable construction. These countries demonstrate a general need for robust regulatory frameworks and enforcement mechanisms to promote and implement sustainable practices. Additionally, a preference for traditional construction practices and a broad lack of awareness about sustainable construction principles are significant barriers. Economic factors, particularly the perception of higher costs associated with green construction, also play a crucial role, indicating a need for more transparent communication about the long-term economic benefits of sustainable practice.

In South-eastern Asia, countries like Malaysia, Indonesia, and Vietnam encounter distinctive challenges in sustainable construction, notably a lack of technical expertise and skilled labour, alongside economic and material limitations that underscore issues related to financing and access to sustainable projects. The critical role of government in this region is highlighted, with a pressing need for more substantial incentives and policies to encourage sustainable practices, indicating a widespread acknowledgement of the importance of governmental action in fostering the adoption of these practices.

Southern Asian countries like Pakistan, Iran, and India present a different set of challenges where cultural attitudes and social norms significantly influence the adoption of sustainable construction. Economic constraints, poverty, and the high perceived upfront costs of green construction are significant

Table 2.1 Barriers encountered in the Asian sustainable construction practices

Barriers in practices		*Central Asia*	*Eastern Asia*	*South-eastern Asia*			*Southern Asia*			*Western Asia*			*Frequency*
		Kazakhstan	*China*	*Indonesia*	*Malaysia*	*Vietnam*	*India*	*Iran*	*Pakistan*	*Kuwait*	*Oman*	*Qatar*	
Economic barriers	Lack of incentives	✔	✔		✔	✔							4/11
	High costs	✔	✔	✔	✔				✔	✔	✔	✔	8/11
	Lack of financing						✔		✔		✔	✔	4/11
Social and cultural barriers	Resistance to change	✔			✔		✔	✔	✔				5/11
	Low demand				✔				✔		✔		3/11
	Attitudinal resistance	✔	✔	✔			✔			✔	✔		6/11
	Traditional practices		✔			✔							2/11
Political and regulatory barriers	Lack of government enforcement	✔							✔	✔	✔		4/11
	Need for policy improvement	✔							✔				2/11
	Poor enforcement	✔	✔	✔							✔		4/11
Technical and knowledge barriers	Need for awareness		✔		✔		✔	✔		✔	✔		6/11
	Lack of communication and collaboration		✔			✔		✔	✔				4/11
	Lack of management support	✔		✔		✔				✔		✔	5/11
	Lack of innovative technology		✔			✔	✔	✔	✔				5/11
	Need for stakeholder understanding	✔	✔		✔							✔	4/11
	Limited knowledge	✔				✔		✔					3/11

barriers in these regions. Furthermore, low literacy rates concerning environmental issues and a general lack of education and awareness about sustainable construction highlight the need for targeted educational initiatives and public awareness campaigns. While there are common barriers across these regions, such as economic concerns and the need for government intervention, each region also exhibits specific challenges that necessitate localized strategies.

With Kazakhstan as an example, Central Asia's journey towards sustainable construction is hindered by a constellation of challenges spanning economic, social, regulatory, managerial, and educational domains. Economic barriers, such as the lack of incentives and high initial costs, combine with social and cultural resistance to change, deeply ingrained attitudes, and a reluctance to adopt new practices. Regulatory obstacles are marked by insufficient policies and weak enforcement mechanisms, further exacerbated by a notable lack of managerial support and commitment towards sustainability within the construction sector. Moreover, a significant educational gap and a general lack of stakeholder understanding impede the widespread adoption of sustainable construction principles. Addressing these multifaceted barriers demands a comprehensive strategy that includes economic incentives, educational campaigns, regulatory reforms, and initiatives to shift cultural perceptions and organizational behaviours. Such a holistic approach is essential for Pakistan to overcome the existing hurdles and embrace sustainable construction practices, contributing to the country's environmental sustainability and economic development.

China, representing Eastern Asia, faces its unique set of challenges, including a need for government incentives for GCTs and extra costs associated with these technologies. In China, significant barriers hinder the widespread adoption of GCTs. The entrenched reliance on traditional construction methods poses a significant obstacle, as does the additional financial burden these advanced technologies often entail. Compounding these challenges is a notable need for more personnel skilled in modern, sustainable construction practices, limiting the capacity to innovate and implement new technologies effectively. Furthermore, entrenched conflicts of interest among construction stakeholders create resistance to change and slow the adoption of more sustainable practices. These barriers are symptomatic of broader systemic issues that require context-specific solutions sensitive to China's unique economic, cultural, and regulatory landscapes. To overcome these hurdles, targeted policy interventions are essential, yet the need for more government incentives for adopting GCTs further complicates the pathway towards sustainability in China's construction industry. Addressing these challenges will involve enhancing incentive structures and improving training and education in sustainable practices among professionals in the construction sector.

2.4 Implemented promotions of sustainable construction in Asia

Table 2.2 is a critical summary of the comprehensive solution strategies employed to tackle the challenges of sustainable construction across Asia.

This table delineates various approaches, organized under several vital categories: political advocacy and government, regulatory frameworks, economic incentives, educational and awareness efforts, technical innovation and practices, and cultural and social engagement. Each category encapsulates specific actions and policies that various Asian countries have implemented, highlighting the multifaceted nature of addressing sustainable construction effectively. It summarizes in-depth explorations of scalable strategies, sustainable technologies, and best practices that have been implemented in the Asian construction industries to promote environmental stewardship, resource efficiency, and energy conservation in construction projects. Table 2.2 serves as a pivotal reference point for stakeholders who aim to foster environmentally sustainable and socially responsible construction practices, driving the industry towards a more sustainable and resilient future.

2.4.1 Political advocacy and governance

The role of government policies and incentives is paramount in promoting sustainable construction. There is a glaring need for robust regulatory frameworks and supportive government initiatives across various Asian countries. Effective government intervention can catalyse the shift towards sustainable practices by setting standards, providing incentives, and ensuring compliance. This includes the development of green construction codes, environmental impact assessments, and policies that encourage the use of sustainable materials and technologies. Government incentives such as tax breaks, subsidies, and grants for green projects can significantly lower the barriers to entry for sustainable construction.

The essence of fostering political will for green initiatives underlines the necessity for governmental leadership and commitment towards sustainable construction practices. This involves not just the endorsement of sustainable projects but also the active participation of government bodies in setting ambitious environmental goals and allocating resources to achieve them. A prime example is Singapore's Green Plan 2030, an integrated national movement to propel the country towards sustainable development. Through strong political will, Singapore has set comprehensive targets for reducing carbon emissions, enhancing green spaces, and promoting sustainable living, effectively positioning itself as a global leader in green building initiatives. The government's commitment is further demonstrated by its investment in green technologies and implementing the Building and Construction Authority's Green Mark scheme. This benchmarking scheme incorporates internationally recognized best practices in environmental construction and performance.

Establishing cross-departmental collaborations for sustainability signifies the importance of synergy between various governmental departments and agencies in orchestrating and implementing sustainable construction policies. By fostering a collaborative approach, governments can ensure that sustainability principles are integrated across all urban planning, construction, and environmental management sectors. An illustrative case is the collaboration

Table 2.2 Solutions to promote sustainable construction practices in Asia

Aspects	*Solutions*
Political advocacy and governance:	Fostering political will for green initiatives
	Establishing cross-departmental collaborations for sustainability
	Providing political backing for international sustainability agreements
	Integrating sustainability into national development plans
Regulatory frameworks:	Introducing mandatory green building codes
	Implementing performance-based environmental regulations
	Establishing compliance mechanisms for green standards
	Creating zoning laws that encourage sustainable development
Economic incentives	Offering grants for sustainable construction research
	Providing low-interest loans for green building projects
	Implementing rebate programs for energy-efficient components
	Designing financial models to support sustainability investments
Educational and awareness efforts:	Launching national awareness campaigns on sustainability benefits
	Developing curriculum on sustainable construction for schools and universities
	Organizing workshops and seminars for developers, builders, and others
	Creating online resources and toolkits for sustainable construction practices
Technical innovation and practices:	Supporting R&D in sustainable materials and construction methods
	Encouraging the use of renewable energy sources in construction projects
	Facilitating the adoption of green building standards and certification
	Promoting the integration of smart technologies for energy efficiency
	Advancing water conservation techniques
	Enhancing material efficiency and waste reduction
Cultural and social engagement	Leveraging traditional, sustainable building practices
	Promoting educational campaigns to shift cultural perceptions towards sustainability
	Showcasing sustainable projects to enhance public engagement
	Involving communities in sustainable development through participatory design

between the Ministry of Urban Development and the Ministry of Environment, Forest, and Climate Change in India, which has led to the formulation of the Smart Cities Mission. This initiative aims to promote cities that provide core infrastructure, offer their citizens a decent quality of life, and adopt sustainable and inclusive development practices. The collaboration has facilitated the creation of green buildings, efficient waste management systems, and renewable energy projects, showcasing how inter-departmental cooperation can significantly enhance the effectiveness of sustainable construction efforts.

2.4.2 Regulatory frameworks

Introducing mandatory sustainable construction systems is a pivotal regulatory framework that mandates specific standards for energy efficiency, water conservation, and environmental sustainability in construction projects. For example, China introduced a Code for green construction of buildings (GB/T 50905–2014) in 2014, to save resources, protect the environment, and ensure WHS of construction workers in construction, expansion, renovation, and demolition projects. The code provides detailed clauses for how to organize and manage green construction from construction preparation, construction sites, footings and foundations, structure, interior decoration, insulation and waterproofing, mechanical and electrical installation, and demolition engineering. Particularly, in China, the Green Construction Evaluation Standards for Building and Infrastructure Projects (GB/T50640–2023) were implemented in 2024 to replace the older vision of Green Construction Evaluation Standards for Buildings implemented in 2010. The new evaluation standard contents mainly include evaluation organization, planning, management, method, and evaluation framework index for environmental protection, resource-saving, human-resource saving, and technical innovation.

Implementing performance-based environmental regulations involves establishing standards that focus on the outcomes rather than prescribing specific methods for achieving sustainability in construction. This approach allows for greater flexibility and innovation, as builders and developers can choose the most cost-effective or technologically advanced methods to meet environmental performance targets. South Korea's Green Building Certification System, often called Green Standard for Energy and Environmental Design, exemplifies this approach by evaluating buildings' overall environmental performance, including energy savings, resource efficiency, and indoor environmental quality. G-SEED encourages builders to pursue innovative and diverse methods to enhance their buildings' sustainability, fostering a competitive environment where green innovation thrives. Through performance-based regulations, South Korea effectively drives its construction industry towards higher environmental standards, contributing to the country's sustainability goals.

2.4.3 Economic incentives

The economic aspect of sustainable construction emerges as a pivotal challenge across Asian countries. A key barrier is the perception of high initial costs associated with green building, which requires stakeholders to embrace sustainable practices. This concern underscores the importance of creating compelling financial incentives and transparently communicating the long-term economic benefits of sustainable construction. Addressing economic barriers requires innovative financing models, such as tax incentives, subsidies, or low-interest loans, to alleviate the upfront costs. Additionally, emphasizing long-term cost savings through energy efficiency, reduced waste, and lower operational expenses can alter the perceived value proposition of sustainable practices. Economic considerations also extend to the impact of green buildings on property values, the potential for job creation in new green industries, and the overall contribution to a country's economic resilience. Integrating sustainable practices into the mainstream construction industry can stimulate economic growth by fostering new markets and industries centred around green materials and technologies. Therefore, economic strategies for promoting sustainable construction should be multifaceted, addressing the immediate financial concerns and the broader economic implications of transitioning to more sustainable building practices.

Providing low-interest loans for green building projects is another economic incentive to lower the financial barriers to sustainable construction. These loans make it more feasible for developers and homeowners to invest in energy-efficient, environmentally friendly buildings by reducing the cost of borrowing. A notable implementation of this strategy is seen in Japan with the "Flat 35S" loan program. The program offers long-term, fixed-rate mortgages at reduced interest rates for homes that meet specific energy-saving standards. By making green homes more affordable, the program incentivizes builders and buyers to prioritize sustainability, driving the demand for sustainable construction. This approach promotes energy efficiency, reduces greenhouse gas emissions, and supports the country's broader environmental objectives while stimulating its green economy.

2.4.4 Educational and awareness efforts

Launching national awareness campaigns and education on sustainable construction is crucial for the industry and stakeholders to understand the importance of sustainable construction and the built environment. These campaigns and educations aim to increase general knowledge about green buildings' environmental, economic, and health benefits, thereby fostering a culture of sustainability. They will engage the public, professionals, and policymakers through workshops, seminars, and media coverage, highlighting success stories and the tangible benefits of sustainable construction practices. An exemplary case is the "Green Building Week" organized by the World Green Building Council. This global event is observed in various countries, providing a platform for sharing knowledge, ideas, and innovations in green building. By raising awareness, such initiatives promote the adoption of green building standards and encourage individuals and businesses to make more environmentally responsible choices.

Integrating sustainable construction principles into the educational curricula of schools and universities is another critical factor in promoting sustainable development in the construction sector. This approach provides students with the knowledge and skills required to innovate and implement sustainable construction practices. In Asia, Singapore stands out for its commitment to incorporating sustainability into its education system. For example, the Singapore University of Technology and Design (SUTD) offers comprehensive courses on sustainable construction and environmental engineering, preparing students to address the challenges of climate change and sustainability in the built environment. Through projects, research opportunities, and collaborations with industry leaders in green building, SUTD ensures that its graduates are well-equipped to contribute to Singapore's vision of becoming a super-low-energy city. By prioritizing sustainability in education, Singapore is cultivating a workforce capable of driving sustainable development and innovation in construction and beyond, setting a benchmark for educational institutions worldwide.

2.4.5 Technical innovation and practices

Supporting R&D in sustainable materials and construction methods is fundamental to advancing technical innovation within the sustainable construction sector. This support facilitates the discovery and implementation of new, eco-friendly materials and energy-efficient building techniques that can significantly reduce the environmental impact of construction projects. South Korea exemplifies a commitment to this approach in Asia through its robust support for R&D in green building technologies. The Korea Institute of Civil Engineering and Building Technology is pivotal in researching high-performance materials, renewable energy integration, and sustainable urban development strategies. Through government funding and partnerships with the private sector, South Korea has developed innovative solutions such as the Eco-Block, made from recycled materials, and green roof technologies that contribute to energy conservation and urban biodiversity. These initiatives demonstrate the country's dedication to sustainable construction and position it as a leader in green technology R&D, inspiring further advancements in the field globally.

Integrating intelligent technologies into construction practices represents a significant stride towards achieving energy efficiency and sustainability in the built environment. Innovative technologies, including digital construction technologies, IoT (Internet of Things) devices, and AI-driven energy management systems, enable buildings to optimize energy use, reduce waste, and improve overall environmental performance. China's push towards intelligent, green buildings showcases the potential of these technologies. The construction project named Sino-Singapore Tianjin Eco-City is a prime example, where intelligent systems control lighting, heating, and cooling to minimize energy consumption while ensuring occupant comfort. The project incorporates renewable energy sources, water recycling systems, and sustainable transportation solutions, embodying the principles of holistic sustainable urban development. By adopting intelligent technologies, China is making significant progress in reducing the carbon footprint of its building sector, setting a standard for future sustainable construction projects across Asia and beyond.

Finally, offering grants for sustainable construction research is a critical economic incentive to encourage innovation in green building technologies and practices. These grants provide financial support for research into new materials, construction methods, and technologies that can reduce buildings' environmental impact. By facilitating such initiatives, countries not only address immediate financial concerns but also support long-term sustainability goals.

2.4.6 Cultural and social engagement

Cultural and social factors play a significant role in adopting and implementing sustainable construction practices. It can be noted that cultural norms and traditional construction methods are presenting substantial resistance to new, sustainable methodologies in Asia. Overcoming this barrier requires a sensitive approach that respects and integrates local cultural values and practices. First, education and awareness campaigns are vital in altering the perceptions and attitudes towards sustainability, emphasizing the long-term environmental, economic, and social benefits. This includes respecting traditional knowledge and practices while introducing sustainable innovations that provide opportunities for skill development and employment. Second, it is important to highlight successful sustainable construction practices and research through exhibitions, open houses, and online platforms. For example, engage artists and designers to create installations and artworks that raise awareness about sustainable construction. This can capture people's attention and spark conversations to see and experience the benefits of sustainable design and construction first-hand. The third way is to incorporate traditional and cultural elements into sustainable construction practices. It requires identifying traditional building techniques, materials, and designs that are environmentally friendly and culturally significant, and then promoting them in practice. Fourth, addressing social dynamics, such as poverty and economic instability, is crucial. Work with policymakers and advocate for regulations and incentives to make sustainable options more accessible to poor communities and public projects. This can include tax benefits, financial support mechanisms, green building certifications, energy efficiency standards, and affordable green solutions. Finally, integrating sustainable construction practices requires a harmonious balance of technological innovation, economic feasibility, and cultural sensitivity, ensuring that these practices are not only environmentally beneficial but also socially inclusive and culturally appropriate.

Particularly, encouraging community-led sustainable construction projects empowers local communities to develop and implement sustainable building practices actively. Engaging communities in the planning and implementation of sustainable projects can foster wider acceptance and participation. Social initiatives like community-engaged projects and participatory design processes can help bridge the gap between modern sustainable requirements and traditional construction practices. This can help create a sense of ownership and pride that align with local needs and values in sustainable development. An exemplary case is in Thailand, where the Community Organizations Development Institute supports community-led housing projects emphasizing sustainability. One notable project involves the Baan Mankong Collective Housing Program, which focuses on secure housing solutions that are affordable, community-designed, and environmentally friendly. These projects often incorporate local materials, community waste management systems, and shared green spaces, promoting environmental sustainability, social cohesion, and resilience. Through such initiatives, Thailand demonstrates how cultural and social engagement can significantly enhance the sustainability of construction projects while strengthening community bonds.

2.5 Conclusion

Sustainable construction practices across Asian countries have significantly contributed to environmental conservation, economic development, and social well-being, showcasing a region-wide commitment to sustainability and innovation. Within the diverse landscapes of Asia, countries from different regions have implemented varied strategies tailored to their unique environmental, cultural, and economic contexts. For instance, South-eastern Asia, Indonesia, and Malaysia have been at the forefront of integrating green building standards and eco-friendly materials into their construction projects, demonstrating a solid commitment to reducing carbon footprints and enhancing energy efficiency. Meanwhile, in the Middle East, countries like Oman and Qatar have made remarkable strides in adopting sustainable urban planning and renewable energy technologies, setting new benchmarks for sustainable development in a region traditionally known for its reliance on fossil fuels. These examples underline the comprehensive efforts and significant contributions of Asian countries to the global agenda of sustainable construction, reflecting a collective move towards a more sustainable and resilient future.

Despite the significant strides made by Asian countries towards sustainable construction, many barriers continue to impede progress across the region. These obstacles vary in intensity and nature from country to country and share common themes that underscore the challenges of transitioning towards greener building practices. Among these, economic barriers emerge as predominant hindrances, including the perception of high costs associated with green construction and a lack of financial incentives. This economic perspective often overshadows the long-term benefits of sustainability, such as energy savings and reduced environmental impact, deterring investment in green construction initiatives. Moreover, regulatory and policy challenges, characterized by insufficient government enforcement, weak regulatory frameworks, and a need for comprehensive policy improvement, further stifle the adoption of sustainable practices. These issues are not isolated but are frequently reported across regions, from the Middle East to Central and South-eastern Asia, indicating a widespread need for more robust regulatory support and effective policy mechanisms.

In addition to economic and regulatory barriers, social and cultural resistance to change poses a significant obstacle to the widespread adoption of sustainable construction practices. Deeply ingrained traditional construction methods and a general lack of awareness about the benefits of green building contribute to this resistance, making it difficult to shift attitudes and behaviours. The challenge is exacerbated by a lack of technical expertise and skilled labour in sustainable construction, particularly in South-eastern Asia. This skill gap hinders the effective implementation of sustainable technologies and practices, emphasizing the need for targeted educational programs and professional training to build capacity in the construction industry. Furthermore, as seen in countries like Pakistan and India, inadequate stakeholder collaboration and a lack of management support underscore the importance of fostering a collaborative environment where stakeholders can work together towards common sustainability goals.

These barriers, from economic and regulatory challenges to social resistance and technical limitations, highlight Asia's complex landscape of sustainable construction. While there is a clear recognition of the need for sustainable development and numerous initiatives have been undertaken, overcoming these hurdles requires a concerted effort. Addressing these challenges will necessitate a multi-faceted approach that includes policy reforms and economic incentives, educational initiatives, stakeholder engagement, and a cultural shift towards sustainability. Only through such comprehensive strategies can Asian countries fully realize the potential of sustainable construction and contribute to a more sustainable global future.

Nuanced promotional strategies tailored to address the region's diverse challenges and opportunities are essential to accelerating the transition towards sustainable construction across Asia, which play distinct roles in catalysing sustainable development within the construction sector. These strategies can be framed as three pillars including foundational, "pull", and "push" actions.

Foundational actions are pivotal, as they set the groundwork for sustainable construction practices to flourish. Central to this foundation is the role of political advocacy and governance. Governments are expected to lead by example, embedding sustainability into national development plans and fostering political will through comprehensive policies and cross-departmental collaborations. Implementing such a strategy requires a concerted effort to integrate sustainability goals into all facets of governance, thereby creating a robust policy environment that supports green initiatives. Another cornerstone is establishing stringent regulatory frameworks, such as mandatory green construction codes and performance-based environmental regulations. These regulations must be enforceable, with precise compliance mechanisms and incentives for adherence, ensuring that sustainable construction becomes the standard rather than the exception.

"Pull" actions provide the tractions for adopting sustainable construction practices, including educational and awareness efforts and promotions, like grants, low-interest loans, and rebate programs, to address the critical barrier of perceived high costs associated with green building and construction. By demonstrating the long-term economic benefits and making sustainable options financially viable, these incentives condition the market to favour green construction. Concurrently, educational efforts are vital in changing perceptions and increasing demand. Developing curricula on sustainable construction, launching national awareness campaigns, and providing resources and toolkits empower stakeholders with knowledge, altering the industry's landscape from the ground up.

"Push" actions are the driving force that encourages stakeholders to engage with and prioritize sustainable construction. This includes technical innovation and practices, such as promoting R&D in sustainable materials and renewable energy sources, which inspire industry advancement and competitiveness. Additionally, cultural and social engagement is a motivational tool, leveraging traditional practices and emphasizing community-led projects and health benefits to foster a more profound, intrinsic motivation towards sustainability. Recognizing and rewarding sustainable construction achievements through certifications and awards can motivate stakeholders to pursue green building practices.

Implementing these solutions requires a multi-faceted approach for overcoming barriers and promoting sustainable construction, combining policy reform, financial modelling, educational programming, and community engagement. By systematically addressing these aspects, Asian countries can pave the way for a more sustainable and environmentally responsible construction industry, significantly contributing to global sustainability goals. However, the implemented promotions primarily focus on industry enhancement and management. In this book, the "push" actions aim to improve the sustainability of project construction, specifically from the perspectives of project management, construction technologies, on-site construction techniques, and sustainable construction materials.

References

Ahmed, A. M., Sayed, W., Asran, A., & Nosier, I. (2023). Identifying barriers to the implementation and development of sustainable construction. *International Journal of Construction Management*, *23*(8), 1277–1288.

AlSanad, S. (2015). Awareness, drivers, actions, and barriers of sustainable construction in Kuwait. *Procedia Engineering*, *118*, 969–983. https://doi.org/10.1016/j.proeng.2015.08.538.

Arif, M., Egbu, C., Haleem, A., Kulonda, D., & Khalfan, M. (2009). State of green construction in India: Drivers and challenges. *Journal of Engineering, Design and Technology*, *7*(2), 223–234. https://doi.org/10.1108/17260530910975005.

Fathalizadeh, A., Hosseini, M. R., Silvius, A. J. G., Rahimian, A., Martek, I., & Edwards, D. J. (2021). Barriers impeding sustainable project management: A Social Network Analysis of the Iranian construction sector. *Journal of Cleaner Production*, *318*, 128405. https://doi.org/10.1016/j.jclepro.2021.128405.

Iqbal, M., Ma, J., Ahmad, N., Hussain, K., Usmani, M. S., & Ahmad, M. (2021). Sustainable construction through energy management practices in developing economies: An analysis of barriers in the construction sector. *Environmental Science and Pollution Research, 28*(26), 34793–34823. https://doi.org/10.1007/s11356-021-12917-7.

Pham, H., Kim, S.-Y., & Luu, T.-V. (2019). Managerial perceptions on barriers to sustainable construction in developing countries: Vietnam case. *Environment, Development and Sustainability*. https://doi.org/10.1007/s10668-019-00331-6.

Saharuddin, S., Hassan, N. F., & Mohd Kamar, I. F. (2022). Barriers in implementing sustainable construction among contractor. *International Journal of Academic Research in Business and Social Sciences, 12*(8). https://doi.org/10.6007/ijarbss/v12-i8/14485.

Saleh, M. S., & Alalouch, C. (2015). Towards sustainable construction in Oman: Challenges & opportunities. *Procedia Engineering, 118*, 177–184. https://doi.org/10.1016/j.proeng.2015.08.416.

Susanti, B., Filestre, S. F. H., & Juliantina, I. (2019). The analysis of barriers for implementation of sustainable construction in Indonesia. *IOP Conference Series: Earth and Environmental Science, 396*, 012033. https://doi.org/10.1088/1755-1315/396/1/012033.

Tokbolat, S., Karaca, F., Durdyev, S., & Calay, R. K. (2019). Construction professionals' perspectives on drivers and barriers of sustainable construction. *Environment, Development and Sustainability*. https://doi.org/10.1007/s10668-019-00388-3.

Wang, Y., Chong, D., & Liu, X. (2021). Evaluating the critical barriers to green construction technologies adoption in China. *Sustainability, 13*(12), 6510. https://doi.org/10.3390/su13126510.

Wong, S. Y., Low, W. W., Wong, K. S., & Tai, Y. H. (2021). Barriers for green building implementation in Malaysian construction industry. *IOP Conference Series: Materials Science and Engineering, 1101*(1), 012029. https://doi.org/10.1088/1757-899x/1101/1/012029.

3 Construction project management methods towards sustainable development

3.1 Introduction

As the social awareness of environmental protection and sustainability has increased, there has been increasing global concern for sustainable development and its impact on various industries. The construction industry is no exception, with sustainable construction becoming a prevailing trend. More and more construction companies have begun to prioritize sustainable construction project management in the construction industry, as it addresses the growing concerns about environmental and social sustainability. It is not only to basically achieve project management targets but also to efficiently use resources and protect the environment. The lack of the application of sustainable construction project management will commonly cause waste of resources and generate energy and environmental pollution. In traditional and common construction project management, it is essential to complete the project on time and budget with high-quality and safety conditions. The primary goals of sustainable construction project management are to avoid environmental pollution, decrease the consumption of non-renewable resources, and commit to social sustainability in the construction process. Therefore, by adopting sustainable practices, construction companies can reduce their carbon footprint, conserve resources, and minimize waste generation, thereby contributing to a more sustainable and environmentally friendly society.

The importance of sustainability in construction project management is growing rapidly due to several factors. First, there is a growing awareness of the environmental and social impacts of construction activities. Second, there is a growing demand for environmentally friendly building products and construction methods. Third, sustainable construction aims to minimize the environmental impact of construction activities while ensuring that they are also economically viable. Finally, there is a growing emphasis on corporate social responsibility and sustainability reporting in many organizations. Consequently, construction companies are increasingly adopting sustainable practices and integrating sustainability considerations into their project management processes.

DOI: 10.4324/9781003202660-3

Sustainable construction project management can be understood as the application and practice of SDGs and sub-goals in construction projects. The first important thing of ensuring the effective execution of sustainable construction project management is to manage the project sustainability through its whole life cycle. Considering sustainability for construction procurement and supply provides the assurance and preconditions for sustainable construction. Risk management will be implemented in sustainable construction to control and reduce the risks that potentially hinder and restrain the implementation of sustainable construction. Accordingly, life cycle management in sustainable construction, sustainable procurement in construction, and risk management for sustainable construction are discussed in this chapter. Moreover, an advanced construction project management method called "lean construction" will be introduced to promote sustainable construction. The approach of Integrated Project Delivery (IPD) will be explored to cooperate and integrate key stakeholders in developing and optimizing sustainable construction copes and objectives, as there is a growing emphasis on collaborating with other stakeholders to develop sustainable construction projects that are inclusive and beneficial for all parties involved.

3.2 Life Cycle Management (LCM) in sustainable construction

3.2.1 What is LCM?

With the rapid development of urbanization, the construction industry is the main consumer of resources and energy causing major environmental pollutants. Most sustainable development plans for the construction industry may focus only on a certain stage (such as the stage of building material production, construction, operation and maintenance, building demolition, and disposal), which can lead to the result that there are energy saving and emission reduction in a certain stage, but the life cycle still has negative environmental effects. In this context, the life cycle analysis is applied to systematically evaluate energy conservation and emission reduction programs in the construction industry, so as to reduce the environmental burden of building materials and achieve sustainable development of the construction industry.

LCM is a management method that views the entire life cycle of products, systems, or projects throughout their entire life cycle as a continuous process. In general, the key significant point of implementing LCM is to manage them and achieve their objectives without only considering one stage such as design or production, but with the highlight of considering their requirements and benefits during their whole life. Standing on sustainable development, LCM will maximize its economic, environmental, and social benefits considering the interrelationships and impacts of each stage of its life cycle. Therefore, LCM has been widely applied in various industries, including product manufacturing, system governance, project management, and more, as it generally can help reduce costs, optimize performance, protect the environment, and promote sustainable development.

3.2.2 *Considering sustainability in the construction project life cycle*

Construction products normally have a long physical life, such that buildings can be generally designed with a life of 30, 50, 70, or 100 years depending on different building materials. The whole life cycle of construction products involves a large amount of natural resources and energy input, material consumption, waste generation, and carbon emissions. Therefore, considering and implementing LCM in the construction industry has more important benefits to addressing environmental and economic sustainability concerns. It could reduce the negative environmental impacts and life cycle cost of construction products and achieve the sustainable development of the construction sector. The whole life cycle of construction products starts from the first planning and design, construction, operation and maintenance, to final demolition. Furthermore, the construction stage can be divided into several processes, consisting of bidding, construction planning, pre-construction, execution, monitoring, handover, and warranty, to study how to promote sustainable construction. The whole life cycle is shown in Figure 3.1.

From the sustainable development viewpoint, the main task of the planning stage is to plan the proposed projects, and analyse their sustainable feasibility, particularly in environmental and social aspects, to make the investment

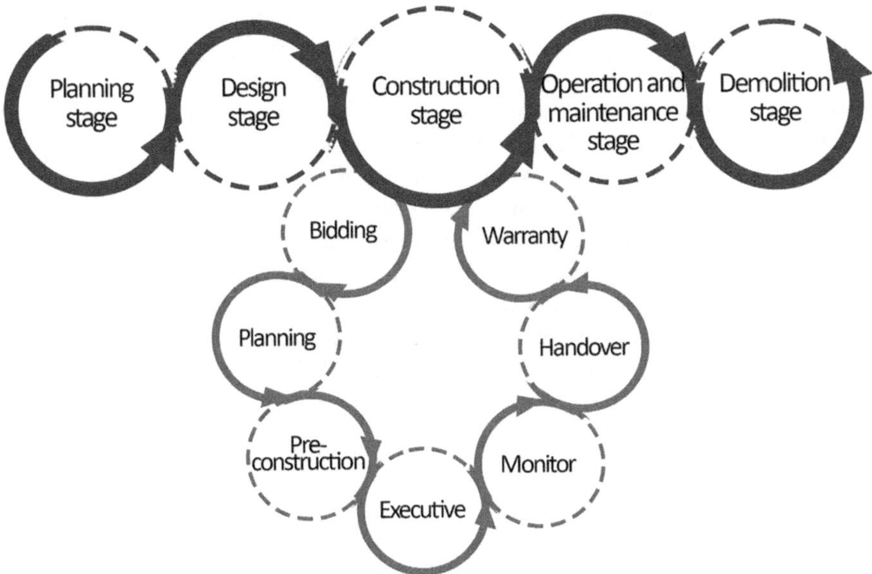

Figure 3.1 The entire life cycle of the project and construction.

decision based on the maximum life cycle return and/or minimum life cycle cost.

The design stage is an important part of construction projects and provides specific design requirements for the implementation of the project. The sustainability of the project should be significantly considered in the design stage. The design information and products have a great impact on the environmental, social, and economic sustainability of the whole life of the project.

The project construction stage is mainly implemented by construction companies and generally includes bidding, construction planning, pre-construction, execution, monitoring, handover, and warranty steps.

- The construction company will decide whether to bid on the project or not, where the company is required to consider whether winning the project will impact the company's sustainability or not. To satisfy the requirements of sustainable procurement, the technical bid document is generally to answer how to achieve the sustainability of the project.
- In the construction planning step, various construction plans are developed, for example, for sustainable construction, environment protection plan, energy management plan, water saving plan, consuming materials and resources plan, construction waste management plan, WHS management plan, socially sustainable construction plan, and risk management plan.
- The sustainable pre-construction works commonly include procuring sustainable materials, using energy-efficient equipment and tools, and laying out environmentally friendly construction sites. Specifically, in this stage, it is important to explicitly and firmly inquire about sustainable construction requirements from all stakeholders, such as through project meetings.
- In the execution step, main construction activities are implemented over a long period. The larger quantities of resources and energy are consumed, and an enormous amount of waste and greenhouse gas (GHG) emissions are generated. The implementation of these sustainable construction plans will provide guidance for construction activities, aiming to save energy and water consumption, reduce waste generation and GHG emissions, and promote WHS and social construction performance.
- Monitoring and controlling the execution of construction activities ensure compliance with sustainable construction regulations, goals, and plans. The monitoring and controlling methods and measures will be addressed to track the construction progress and issues, such as PDCA (Plan-Do-Check-Act), stakeholder management, continuous improvement, and risk management.
- In the handover step, several primary sustainable construction works will be implemented. First, ensure the construction products satisfy the design and sustainable construction goals, such as achieving green building assessment standards and managing site waste recycling and disposal. Second, test and verify the functionality of systems and components, for example, the HVAC (heating, ventilation, and air conditioning) service systems in buildings. Third, provide documentation, certification, training, and support to the owners related to efficiently occupying and operating buildings and other products.

- During the warranty phase, sustainable construction practices remain integral to ensuring the continued effectiveness and longevity of the project. For example, quickly address the remaining defects or sudden issues, particularly if the issue is related to occupant health and safety and waste resources. Moreover, evaluate the current sustainable operation conditions (e.g. energy efficiency and water consumption), and provide suggestions, training and support for adapting new efficient systems for changing conditions. Finally, focusing on maintaining or enhancing sustainability features, develop a maintenance plan and schedule with the owners and other stakeholders, for example, biodiversity and landscaping maintenance.

When the building is put into use, it should maintain the building's conditions and extend the building's lifespan by taking positive operation and maintenance measures, saving costs, reducing environmental impacts, and creating healthy and safe places. Regular maintenance and inspections are required to monitor the building's conditions, particularly the building's HVAC systems. For example, conduct regular energy audits to identify and address inefficiencies, and regularly check water consumption and repair water leaks. Furthermore, it should have significant environmental and economic benefits to use clean and energy-saving technologies such as solar panels or building automation and control systems, to install water-efficient fixtures and implement rainwater harvesting systems for landscape irrigation, and to establish a comprehensive recycling program and a project retrofitting plan to improve the energy efficiency of older buildings.

In the final stage of the building life cycle, there are lots of sustainable responsibilities and works in building demolition. First, establish and implement a comprehensive program to identify, recycle, and reuse demolition material. Particularly, ensure compliance with environmental regulations for the proper disposal of demolition materials. Second, sustainable demolition methods should be applied, such as employing deconstruction demolition methods to disassemble components rather than destroying them. Finally, sustainable demolition processes should be executed and monitored, such as controlling dust, emissions, noise, and waste. In summary, this should be a resource-efficient process of dismantling or tearing down structures while maintaining WHS and minimizing the impact on the environment.

3.2.3 Promoting the LCM method in sustainable construction

Stakeholder collaboration through the entire life cycle

Effective LCM requires collaboration among various stakeholders. Key stakeholders in construction include designers, builders, material suppliers, engineers, developers, building users, project managers and consultants, and local communities. Particularly, the stakeholders will be changed and different along with the progress of project construction in different stages. Understanding and considering the concerns of each stakeholder in different life cycle stages help ensure that construction products and projects are successful, sustainable, and aligned with sustainable goals. More importantly, it is necessary to motivate them to collaborate and promote sustainable construction practices throughout the entire life cycle. For example, inviting end-users and final consumers to engage in project planning and design stages is important to promote sustainable construction practices in the entire life cycle, particularly in the maintenance and operational stages.

Life Cycle Assessment (LCA) for sustainable construction

The ISO14040 standards on LCA were formulated and promulgated in 1997, which defines LCA as the summary and evaluation of energy and material input and output as well as potential environmental impact in the life cycle of a product system. As a tool for the environmental impact analysis of a product, the LCA method plays an important role in assessing cleaner production, product ecological design, waste management, and ecological industry, and provides an important method for analysing the sustainable construction performance of construction projects in the entire life cycle. LCA involves evaluating each stage of projects, from raw material acquisition, manufacturing/construction, use/operation, to waste disposal, considering their impact on the environment and society. In the context of sustainable construction, it typically includes several steps:

- Data collection: Collecting and identifying data related to sustainable construction practices in the entire life cycle.
- Data analysis: Analysing and calculating the results of sustainable construction performance through various tools and techniques such as descriptive statistics, regression analysis, exploratory data analysis, principal component analysis, and time series analysis.
- Performance evaluation: Evaluating and discussing the sustainable construction performance standing on the life cycle viewpoint based on assessment methods such as impact assessment, risk assessment, network analysis, and driving force analysis, and so on.
- Developing, implementing, assessing, and managing improvement strategies: Based on the evaluation results, developing sustainable, effective, and practical improvement measures to improve sustainable performance and reduce negative environmental impacts based on evaluation results. PDCA cycle and other continuous improvement methods may be used to implement and control these strategies.

Life Cycle Cost Management (LCCM)

LCCM in sustainable construction involves considering the total cost of a building or infrastructure project over its entire life cycle, from initial planning and design to construction, operation, maintenance, and potential

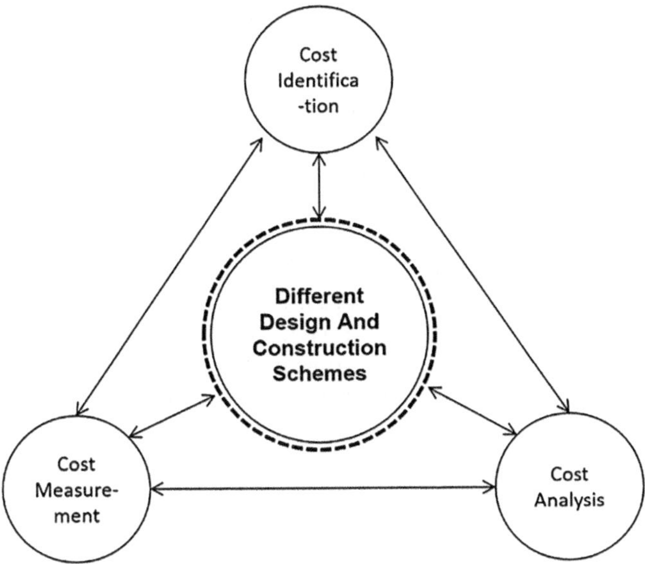

Figure 3.2 The key steps of LCCM.

decommissioning or renovation. Life cycle cost measurement goes beyond initial construction costs and considers ongoing operational and maintenance expenses, as well as potential end-of-life costs. LCCM aims to make informed decisions that optimize life cycle costs while considering factors such as sustainability, durability, and long-term performance. Therefore, this approach can not only help decision-makers to optimize project cost from an entire life cycle point but also benefit from minimizing the environmental impact of construction by considering environmental costs. The key steps of LCCM are shown in Figure 3.2. Here are the key steps and strategies for implementing LCCM to assess and manage different sustainable construction practices.

- Cost identification: This work generally happens in the planning and design stages. It is to identify all types of costs in all stages, and consolidate them into a life cycle cost accounting table. It is important to consider and collect all cost information related to different construction materials, construction methods, building service systems, technologies, and design options, such as the installation, operation and maintenance costs of the HVAC systems. The amount of GHG emissions can be estimated particularly in the construction, operation, and demolition stages, which should include the embodied carbon emissions of all consumed resources.
- Cost measurement: Different feasible and acceptable project design and construction schemes will be developed, based on different materials, energy-efficient systems, and technologies. The total costs of these different

schemes will be separately calculated from the entire life cycle. The time value of these costs should be considered. The cost of GHG emissions could be calculated based on carbon tax and trade price.

- Cost analysis: Based on the results of cost measurements, decision-makers can analyse different design and construction schemes and find the best sustainable construction scheme. Different decision-makers may choose different schemes. However, under the same conditions and constraints, it should be encouraged to select the minimum life cycle cost, not the minimum investment cost or construction cost. Transparent disclosure and analysis of costs are generated in specific stages. More importantly, the scheme with the materials, components, and systems that can be mostly recycled and reused is preferred. The value-engineering approach is applied to measure the effect of sustainable construction and to conduct cost analysis, including three basic elements: value, function, and life cycle cost, with the lowest life cycle cost to achieve the most effective performance. It provides a scientific standard for identifying cost-effective alternatives without compromising performance or quality, and for optimizing the ratio of sustainability and functionality to cost over the life cycle.

In summary, effective LCCM is an indispensable approach for sustainable construction. Sustainable construction efforts must focus on methods and techniques from inception to demolition in the entire project life cycle, emphasize stakeholder collaborative efforts and integrated processes, support the approaches recommended by LCA, and target the best design and construction scheme with the minimum life cycle cost.

3.3 Sustainable procurement in construction

3.3.1 What is sustainable procurement in construction?

The core characteristic of sustainable procurement (SP) is that it simultaneously considers economic, environmental, and social factors throughout the procurement process. The UK government's report *Procuring the Future* (DEFRA (Department for Environment, Food & Rural Affairs), 2006) defines SP as "using procurement to support wider social, economic, and environmental objectives, in ways that offer real long-term benefits". The British Standard (BS8903) outlines SP objective as follows: "Procuring sustainably allows organizations to meet their needs for goods, services, works, and utilities in a way that achieves value for money on a whole life basis in terms of generating benefits not only to the organization, but also to society and the economy, while minimizing damage to the environment". Accordingly, SP can be defined as the process of acquiring goods, services, works, and utilities in a manner that addresses environmental, social, and economic considerations and objectives simultaneously. Compared to traditional procurement, SP places greater emphasis on considering environmental and social factors, prioritizing suppliers with minimal or no negative impacts on the environment and society.

In the context of construction projects, SP is an essential strategy for achieving sustainability and adapting the built environment to climate change. The procurement process in construction will develop procurement objectives with a critical point for incorporating sustainability objectives. Integrating sustainability into construction procurement involves minimizing resource procurement, reducing negative impacts across the whole procurement and supply chain, ensuring resources are delivered on time and in quality, and considering social responsibilities such as local community engagement and ethical practices. SP supports long-term cost-benefit analysis by considering value for money on a whole-life basis, generating benefits to the organization, society, and the economy while minimizing environmental damage. It is of great significance in the construction industry due to its potential to address economic, social, and environmental challenges and foster sustainable development.

First, by considering environmental factors, SP practices allow the selection of environmentally friendly materials, energy-efficient equipment, and sustainable construction contractors and methods. Therefore, construction projects can minimize their carbon footprint, conserve natural resources, and mitigate climate change impacts.

Second, by implementing SP practices, the buyer and winning bidder display their corporate responsibilities, reputation, and competitiveness in sustainable construction. By considering social factors, construction projects aim to provide fair opportunities to the market. SP also emphasizes transparency and equity in the supply chain, ensuring fair treatment of suppliers and promoting inclusive practices.

Third, SP implementation in construction projects enhances environmental, social, and economic feasibilities, particularly for mega and public projects. From the life cycle viewpoint, adopting green project designs and sustainable construction practices can reduce project life cycle costs and improve long-term sustainable performance. Moreover, incorporating SP in construction can attract investors, government incentives, stakeholder engagement, and public support. These supports are much more important for mega and public projects. As a result, construction projects become more resilient to environmental risks (ERs), reduce maintenance costs, and provide long-term value to owners and users.

Finally, standing the SDGs in the world, SP can directly promote the application and enhancement of sustainable designs, materials, equipment, methods, and technologies in the construction industry, and indirectly promote the innovation and development in producing these sustainable products and implementing these sustainable practices. SP encourages the buyer to procure them and the bidders to provide them aiming to win the project bid. Therefore, SP can stimulate innovation, leading to cost-saving solutions and improved sustainable outcomes.

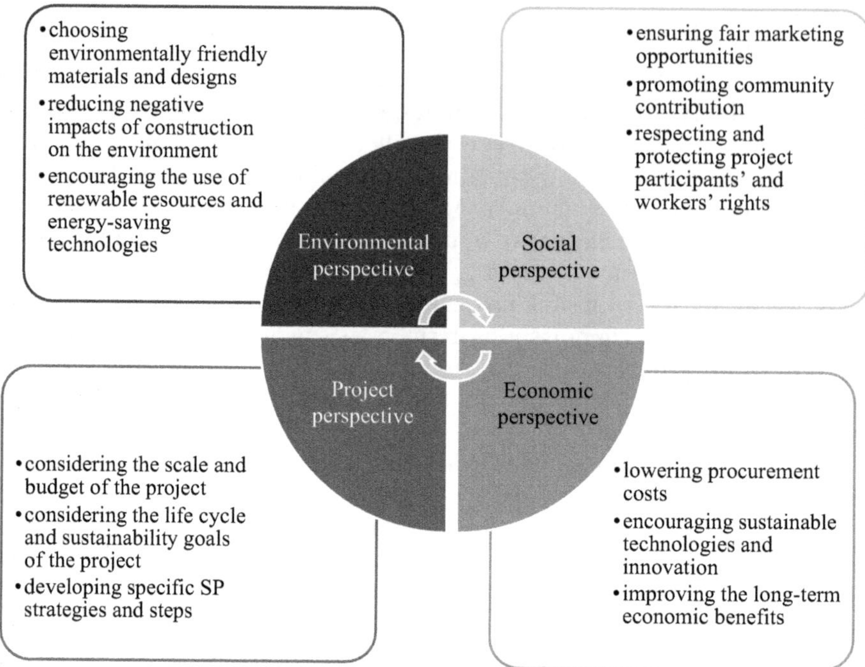

Figure 3.3 Four aspects of sustainable procurement in construction.

3.3.2 Sustainable procurement framework in construction management

SP in construction projects encompasses various aspects, aiming to integrate and satisfy ecological, economic, and social sustainability within the procurement process to achieve project objectives. This involves multiple management elements from different stakeholders' perspectives, typically including four main aspects: environmental, social, economic, and project aspects, as illustrated in Figure 3.3.

From the environmental perspective, procurement focuses on selecting eco-friendly building materials and designs, as well as reducing the environmental impact of the construction process. This can be exemplified using certified sustainable materials, such as the installation of green roofs and walls to mitigate urban heat islands. Furthermore, promoting the use of renewable resources and energy-efficient technologies can significantly decrease a project's carbon footprint.

From the social perspective, procurement emphasizes fair marketing opportunities, community contribution, and the respect and protection of project suppliers' rights. Normally, local suppliers are recommended, due to partnering with local communities and building a more sustainable supply chain. SP also asks the suppliers to adhere to ethical and social sustainability practices.

Moreover, the social aspects of SP prioritize worker safety and fair payment for all project participants.

From an economic perspective, procurement seeks to lower procurement costs while encouraging sustainable technologies and innovation. This can be achieved through life cycle planning and strategic sourcing, as well as leveraging local suppliers who offer competitive pricing and high-quality products. For instance, utilizing durable and low-maintenance materials can reduce long-term maintenance costs and increase the project's overall economic viability.

From the project perspective, SP in construction projects involves considering factors such as project scale, budget, quality, time, and sustainability goals. This requires the development of tailored SP strategies and plans that align with the project's unique context and objectives. For instance, a large-scale project with a limited budget might prioritize cost-effective sustainable solutions, while a smaller project with a strong focus on innovation might prioritize the procurement of cutting-edge sustainable technologies. By integrating sustainability principles into the procurement process, project decision-makers can procure more sustainable products and construct more sustainable projects, fostering a more sustainable future for all.

Ershadi et al. (2021) summarized the required contents of SP in construction projects, standing on the whole life cycle of construction procurement, as shown in Table 3.1.

3.3.3 Promoting sustainable construction procurement

Integrating SP practices leads to the creation of sustainable and resilient construction projects. The successful implementation of SP in construction projects is influenced by various factors, including understanding, policies and strategies, organizational capabilities, and informational barriers. Addressing these constraints is essential for the implementation of SP in the construction industry. The following are some of the key strategies to promote SP in construction.

- Enhance awareness and understanding of SP. A comprehensive understanding of SP, including its policies and strategies, is essential for successful implementation. Increasing awareness and understanding of SP among stakeholders can garner their support and boost the efficiency of SP in construction projects.
- Develop SP policies and strategies. The availability and effectiveness of SP policies and strategies play a crucial role in their successful implementation. Particularly, the government could develop SP policies and strategies for public projects, which could lead to SP implementation in the whole construction industry.
- Improve SP project management skills. The SP project management skills are essential for the successful implementation of SP in construction, such

Table 3.1 The required contents of SP in construction projects (Ershadi et al., 2021)

Stages of the procurement process	*Sustainability requirement*	*Aspects of sustainability*		
		Environmental	*Economic*	*Social*
Pre-procurement decisions	Allocate adequate budget and resources to implement SPM principles.	✓	✓	✓
	Social responsibility and the commitment of senior managers to incorporating sustainability in the procurement process.			✓
Plan procurement	Follow a life cycle analysis to evaluate the environmental friendliness of products and packaging supplied by available suppliers.	✓		
	Ensure endurance, stability, safety, and optimal performance of the equipment to be installed towards delivering a resilient facility/infrastructure.		✓	✓
	Consideration of optimal energy/water consumption and minimum greenhouse gas/toxic emissions in supplying project equipment and material.	✓		
	Avoid noise pollution by supplying equipment that has minimum noise and vibration.	✓		
	Give preference to local suppliers (support the local economy).		✓	
	Give precedence to small suppliers.		✓	
Conduct procurement	Close consideration of design features in procurement to minimize waste.	✓		
	Ensure shortlisted suppliers comply with labour laws and regulations.			✓
	Obligate suppliers to adhere to waste reduction policies.	✓		
	Ensure safe operation and transfer of product to project facilities.			✓
Control procurement	Reduce packaging material.	✓		
	Waste recycling and reuse on construction sites.	✓		
	Awareness of sustainability objectives among people.			✓
Close procurement	Ensure meeting sustainability goals upon completion of the work package.	✓	✓	✓
	Site decontamination and waste removal after contract closure.	✓		

as the professional skills of how to engage the environmental and social sustainable requirements during the whole construction procurement process without decreasing economic feasibility. In the project management team, project manager, procurer, material manager, and quantity surveyors could have the abilities and skills to practise SP.

- Developing sustainable supply chain networks. The availability of suppliers offering sustainable materials, products, and services may be limited, making it difficult to source sustainable options throughout the supply chain. SP involves building supply chain networks and platforms aiming to identify sustainable manufacturers and suppliers. Additionally, regular evaluations of the sustainability performance of manufacturers and suppliers should be conducted to ensure compliance with the company's sustainable standards.
- Applying digital supply chain management technologies. Project stakeholders can use the advanced digital supply chain management technologies to manage the whole construction procurement and supply chain process such as signing contracts considering sustainability, transparently delivery, and ensuring the delivery on time and in quality. Currently, digital technologies mainly consist of blockchain, digital twins, Internet of Things (IoT), artificial intelligence (AI) analysis, robots and automation equipment, and other technologies.

3.3.4 Case study: a practical guide for sustainable procurement in Australia

In 2021, the Commonwealth of Australia published "Sustainable Procurement Guide—A Practical Guide for Commonwealth Entities" to guide Australian agencies to take responsibility for waste through sustainable procurement practices. The guide is a practical resource to help agencies achieve the government commitment that every procurement undertaken by an Australian Government agency will consider environmental sustainability and the use of recycled content when determining value for money. It offers a framework for the Australian Government to enhance sustainability outcomes and integrate sustainability principles into upcoming procurement initiatives.

The Australian government is committed to building a circular economy, particularly to transform waste into a resource. Stable demand for green goods and services through sustainable procurement is the foundation of building a circular economy during the whole supply chain process. The foundation means that the demand provides the initial motivations, such as investment and innovation, to produce these goods and supply these services. Sustainable procurement demonstrates the demands such as waste disposal and the cost of operations and maintenance over the life of the goods and services. Moreover, it asks the procurement decision-makers to make decisions in line with the Australian obligation to spend public money efficiently, effectively, economically, and ethically. Accordingly, the Australian Government has committed to consider environmental sustainability when purchasing goods and services, particularly for products containing recycled content.

This guide provides step-by-step guidance on how to consider sustainability in the different stages of the procurement process, from identifying the business need, procurement actions, the end of the contract, and to reviewing and improving, as in Figure 3.4 (Commonwealth of Australia, 2021). Detailed information relevant to sustainability requirements and procurement actions is displayed in each section of the guide. The guide can be applied within the departmental frameworks of government entities and the Commonwealth Contracting Suite. It is also suitable for officials in non-corporate Commonwealth entities and prescribed corporate Commonwealth entities. The case studies and support tools are provided in the guide, intending to illustrate sustainable procurement for goods and services containing recycled content. The appendices of the guide encompass eco-labels, certifications, standards, and product stewardship schemes. Additionally, they feature model clauses for approach-to-market requirements and contract terms, aiding in the achievement of the desired sustainability outcomes in procurement.

3.4 Risk management for sustainable construction

3.4.1 What is risk management for sustainable construction?

With the development of green building and sustainable construction, there has been an emphasis on risk management in the construction industry. The concept of sustainable construction is rooted in the broader framework of sustainable development, which emphasizes the need for balance between economic, social, and environmental objectives. Compared to traditional construction, sustainable construction generally has more construction objectives and requirements such as environmental and social sustainability, and its construction involves a more complex process and constraints. Accordingly, more risks will be met to implement sustainable construction. It is much more necessary to employ risk management in sustainable construction.

The importance of risk management in sustainable construction can be attributed to several factors. First, construction projects are inherently complex and unpredictable, involving numerous stakeholders, technologies, and environmental factors. These projects often span across multiple disciplines and sectors, making them vulnerable to a wide range of risks that can impact their sustainability. Second, the construction industry is a contributor to environmental degradation and social injustice, particularly when risks are not effectively managed. Consequently, risk management helps to mitigate these negative impacts, promoting social, economic, and environmental sustainability. This involves identifying, assessing, and prioritizing risks across various dimensions such as technical, environmental, social, and economic aspects. Furthermore, it calls for the development of risk management strategies that incorporate the interests of all stakeholders, promote transparency

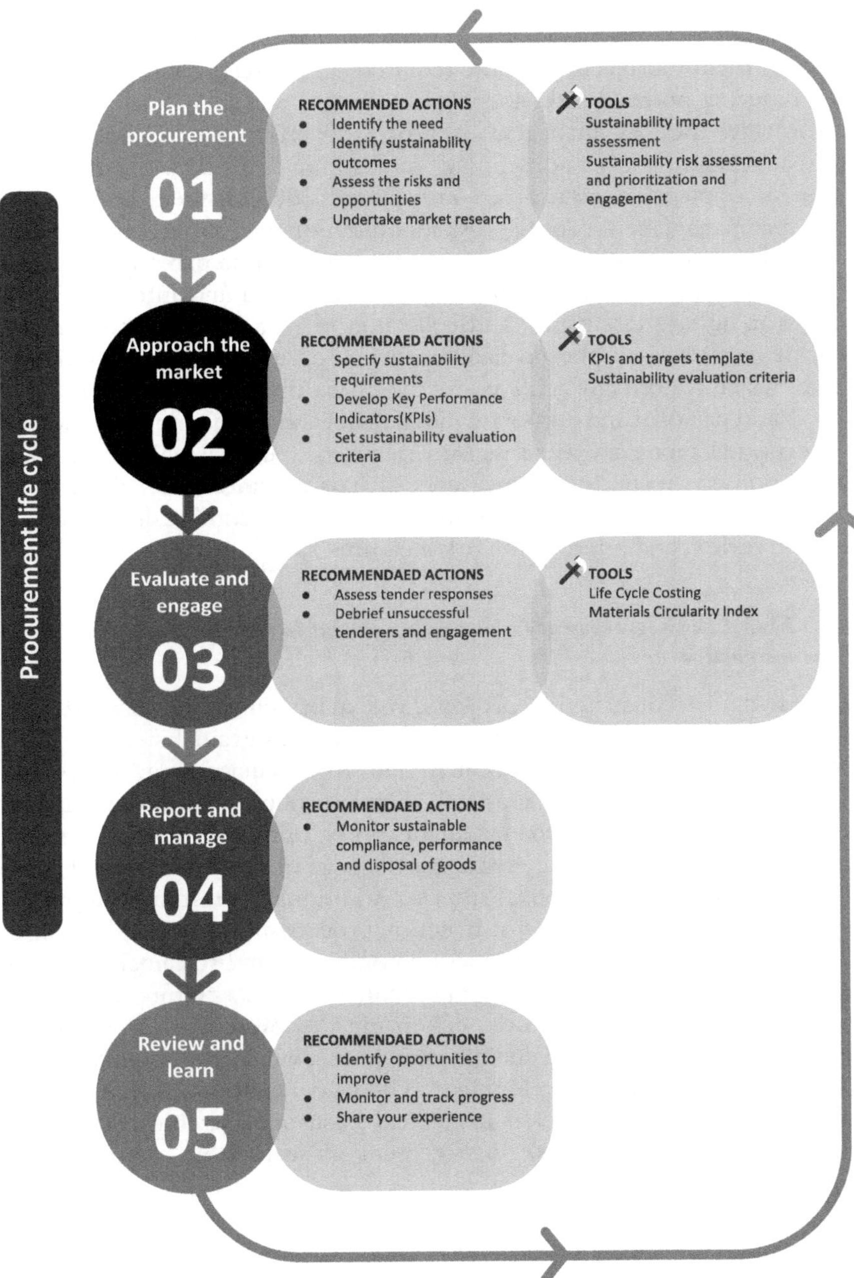

Figure 3.4 The steps of sustainable procurement.

and accountability, and foster long-term sustainability. Particularly, risk management can provide guidance and reference for project teams, helping them to comprehensively identify and manage risks. It is necessary to ensure that construction projects meet the requirements of sustainable development while reducing potential risks and uncertainties.

Normally, the risk management process consists of risk identification, risk assessment, risk response, and monitoring and adjustment. A risk management plan for sustainable construction is developed at the beginning of the risk management process. Risk identification is to identify the risks that could potentially cause hindrance and damage to sustainable construction. The risks can be identified by individual consultation and interview, group brainstorming and discussion, and collection of data and publications. Risk assessment is to assess the probability of risk occurrence and its severity if occurs, which is normally achieved by the qualitative methods (e.g. fuzzy set analysis and personal and corporate judgement) and the quantitative methods (e.g. expected monetary value, sensitivity analysis, and Delphi peer groups). Risk response is to make risk measures such as risk avoidance, risk transfer, and risk reduction. Risk review and monitoring are to continually implement, monitor, review, and improve the risk measures.

3.4.2 Continuous improvement for risk management in sustainable construction

In the realm of construction projects, risk management has emerged as a critical component for ensuring project success and sustainability. To achieve sustainable construction requirements in risk management, continuous improvement is a necessary approach. First, through continuous improvement, the processes of risk management can be optimized to increase accuracy and effectiveness, thereby reducing the risks of sustainable construction and improving the likelihood of success. Additionally, continuous improvement ensures that sustainable construction proceeds smoothly and can adapt timely adjustments to new construction conditions and sustainable requirements. Furthermore, continuous improvement fosters a culture of risk management within the organization, enhancing the awareness and abilities of team members in identifying and managing risks, which lays a foundation for sustainable construction development. Therefore, continuous improvement is necessary to implement risk management in sustainable construction. The process of implementing risk management in sustainable construction is shown in Figure 3.5.

- Development of risk management plans: Clearly define the objectives, scope, and responsibilities of risk management to guide subsequent risk assessment and management processes.
- Training and awareness: Provide comprehensive risk management training to project team members, enhancing their abilities in risk identification,

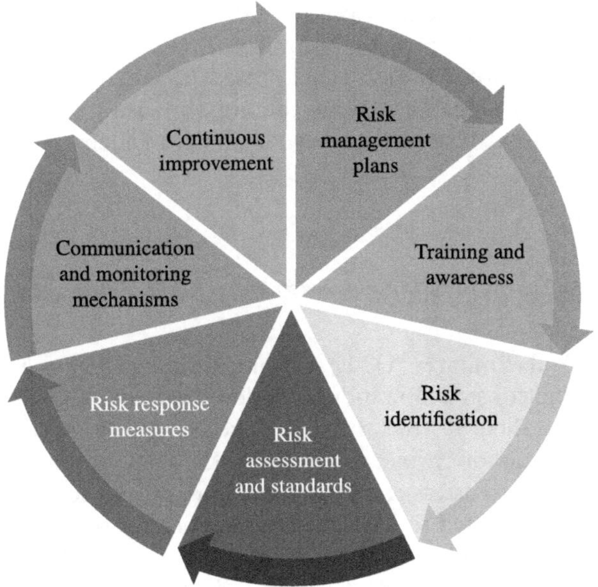

Figure 3.5 Continuous improvement to implement risk management in sustainable construction.

assessment, and response for sustainable construction. Raise general awareness of risk management within the organization.

- Risk identification and the establishment of risk databases: Collect and organize historical project risk information to establish a risk database, facilitating real-time risk assessment. And then identify the potential risks influencing sustainable construction for this project. Continuously update the risk database to have real-time awareness of the risks faced by the project.
- Risk assessment and development of risk assessment standards: Develop specific risk assessment standards and requirements for sustainable construction to ensure the objectivity and accuracy of assessment results. Determine the priority of risks based on severity, occurrence probability, and impact range, providing a basis for formulating risk response strategies.
- Formulation of risk response measures: Tailor risk response measures to different risk levels, ensuring construction sustainability, safety, and quality. Implement proactive preventive measures for high-level and medium-level risks to be acceptable.
- Establishment of communication and monitoring mechanisms: Establish effective risk communication mechanisms to ensure timely sharing of risk information and smooth implementation of response decisions. Continuously monitor changes in risk management during the construction process and adjust risk response strategies accordingly.

- Continuous improvement: Continuously learn from risk management experiences during the project, optimize risk assessment and management processes, and improve project management capabilities. For occurred risk events, conduct root cause analysis, identify the issues, and develop corrective and preventive measures to prevent similar incidents from recurring.

3.4.3 Case studies

Case 1 Risk management in sustainable construction in the United Arab Emirates

In the United Arab Emirates (UAE), sustainable construction has become the mainstream of the construction industry's development. A study by El-Sayegh et al. (2021) identified 30 risks through developing a questionnaire to assess sustainable construction projects in the UAE. The questionnaire used a 5-point Likert scale to assess the possibility, impact, and severity of each risk. The survey was distributed to 200 construction professionals in the UAE with experience in sustainable construction projects, and feedback was received from 44 individuals (a 22% response rate). The Relative Import Index was used to identify and rank these risks. The identified and ranked risks are shown in Table 3.2. The study categorized risks into five main areas:

- Management Risks: These include potential risks encountered during the management of construction projects.
- Technical Risks: These arise during construction and may include design changes, insufficient or inaccurate information on sustainable design, and frequent meetings with green experts and others.
- Risks from the Green Stakeholder Team: This encompasses risks such as client resistance to new green concepts and the limited experience of consultants and contractors.
- Risks Associated with the Use of Green Materials and Technology: These involve dealing with material shortages and handling and quality risks associated with materials.
- Regulatory and Economic Risks: These relate to any regulations or economic crises enforced by the government.

In terms of risk assessment, the top ten risks included four management risks and four technical risks. To address management risks, those in charge need to develop and implement proper project management methods, especially in terms of funding and costs. To overcome technical risks, owners should choose suitable designers with experience in sustainable construction projects to produce designs with minimal changes and reasonable outcomes. For example:

- Shortage of client funds: This could be due to poor planning of sustainable construction costs and schedules.

Table 3.2 The identified and ranked risks in sustainable construction projects in the United Arab Emirates

Categories	*Description*	*Ranking*		
		Probability	*Impact*	*Severity*
Management	Tight schedule for sustainable construction	2	24	4
	Improper sustainable project feasibility and planning	23	27	23
	Shortage of clients' funds	9	3	1
	Difficulty in project budgeting due to unfamiliarity with green projects	6	16	8
	Poor project manager skills related to sustainable construction	14	2	14
	Additional costs due to green material and equipment	1	20	6
	Poor quality of sustainable construction work	19	8	17
Technical	Design changes	3	4	3
	Insufficient or incorrect sustainable design information	4	18	2
	Improper or incomplete green specifications	7	21	7
	Poor scope definition of sustainable construction	5	13	5
	Failure to meet the green code or certification	22	30	20
	The delay caused by frequent meetings with green specialists	15	28	22
	Poor selection of construction techniques in sustainable construction	18	12	15
	Poor productivity of labour and equipment in sustainable construction	13	14	13
Green Team (Stakeholders)	Resistance from the client to adopt new green ideas	24	17	25
	Limited experience as a consultant about sustainable construction practices	11	6	11
	Limited experience of the contractor about sustainable construction practices	12	22	9
	Limited availability and reliability of green subcontractors	10	10	16
	Limited availability and reliability of green suppliers	8	1	10
	Shortage in labourers skilled in sustainable construction	16	23	18
Green Materials and Technology	Poor performance of green materials	30	26	30
	Shortage of green materials	28	25	28
	Long lead time for green materials	20	7	21
	Inappropriate handling and storage of green materials	29	19	24
	Lack of documents and information for new green technologies	27	11	27
Regulatory and Economic	Delay in government approvals for green construction	17	9	12
	Changes in sustainable construction codes and regulations	25	15	26
	Inflation of green materials' prices	21	29	19
	Currency volatility worsened by the import of green materials	26	5	29

Source: El-Sayeg et al. (2021)

- Insufficient or inaccurate information on sustainable design: A lack of experience and knowledge in implementing sustainable design.
- Design changes: A lack of experience and knowledge in implementing sustainable design.
- Unreasonable construction schedule: The UAE's construction industry is known for its rapid pace.
- Unclear definition of the scope of sustainable construction: A lack of experience and knowledge in implementing sustainable construction.

Case 2: Management of environmental risks in highway construction projects (HCPs) in Sri Lanka

Environmental risks (ERs) in construction projects are primary risks in sustainable construction. Abhayantha et al. (2023) investigated the ERs of contactors in HCPs in Sri Lanka to explore how ERs can be managed. A quantitative research approach with three rounds of Delphi was used to analyse and identify the ERs and the measures for managing ERs. Eleven most significant ERs for HCPs were identified and ranked. Twenty-four most appropriate risk response measures were determined, which are mainly the responsibility of contractors in Sri Lanka. The key findings are displayed in Table 3.3. This study will benefit how to manage risks in sustainable construction.

3.5 Sustainable construction through lean construction

3.5.1 *From Lean production to lean construction*

Lean Production originated from Japan's Toyota Motor Corporation. In the 1970s, Taiichi Ohno improved Toyota's delivery time and product quality to the world's leading position through the application of "lean production". The essence of lean production is a production management technique that can significantly reduce idle time, job switching time, poor quality inventory, unqualified suppliers, product development and design cycles, and unqualified performance. The application of lean production in the manufacturing industry has greatly reduced manufacturing costs, shortened the development and manufacturing cycle, and significantly enhanced the competitiveness of the enterprise.

In 1992, Lauri Koskela first proposed the idea of applying "lean thinking" to the construction industry in the report *Application of the New Production Philosophy to Construction* and hosted the first meeting of the International Group on Lean Construction in Finland in 1992. Compared with the manufacturing industry, the product quality and labour efficiency of the construction industry are relatively inefficient. Particularly, the long-term construction process and complex construction environment lead to more construction difficulties and challenges. Each construction product is unique and disposable, but there are repetitions and cycles for the production of

Table 3.3 Primary ERs and response measures in HCP in Sri Lanka (Abhayantha et al., 2023)

Ranked risks	*Most appropriate risk response measures*
Natural resource depletion in local areas such as soil, sand, aggregate, etc.	• Introduce a point system to encourage the use of recycled material. • Educate project stakeholders about circular built environment principles such as reduce, reuse, and recycle in highway construction work. • Design highways by considering the environmental effects more than the cost-effectiveness.
Landscape change and fragmentation	• Strictly adhere to an environmental management plan. • Introducing new forest reservations to compensate for the damages to the environment. • Change the designs with the help of environmental specialists to avoid ERs. • Make the approval process of the environmental impact assessment report very strict by government authorities.
Disturbances to the migration routes of animals	• Provide bio links, animal overpasses, canopy bridges, eco ducts, and underpasses designs. • Strictly adhere to the environmental management plan • Erect meshed fences to avoid the entrance of animals to highways. • Erect sign boards at the places where animals are likely to enter.
Flood	• Keep opening sizes of the bridges and culverts large to allow the flood flow to pass freely. • Mitigate improper drainage and water logging by introducing new designs such as collector drains. • Develop an emergency plan and risk assessment matrix to identify the severity and likelihood of expected risks. • If the current construction activities are affected by unexpected conditions, then go for alternative construction methods.
Loss of biodiversity	• Strictly adhere to the environmental management plan. • Make the EIA report approval process very strict by government authorities. • Maintain the natural drainage network intact to allow for the free movement of the aquatic species. • Design the highway by considering environmental effects more than cost-effectiveness.
Landslides in mountainous areas	• Employment of external specialists in case in-house skills are not available. • Involvement with joint inspections and monitoring mechanisms with government regulatory bodies such as the road development authority. • Develop an emergency plan and risk assessment. • Matrix to identify the severity and likelihood of expected risks. • Design the highway by considering environmental effects more than cost-effectiveness.

(*Continued*)

Table 3.3 (Continued)

Ranked risks	*Most appropriate risk response measures*
Disruption to the physical environment	• Contractors can transfer the losses caused by environmental risk occurrences through insurance companies. • Do the pre-crack survey before the commencement of construction works. • Introduce a crop diversification plan or provide temporary jobs for the victim farmers. • Strictly adhere to the environmental management plan.
Loss of vegetation	• Introducing new forest reservations to compensate for the damages to the environment. • Strictly adhere to the environmental management plan. • Introduce a crop diversification plan or provide temporary jobs for the victim farmers. • Make the EIA report approval process very strict by government authorities.
Impacts on the natural drainage patterns and issues of irrigation water	• Mitigate improper drainage and water logging by introducing new designs such as collector drains. • Maintain the continuity of irrigation and drainage canals/paths within construction sites. • Maintain the natural drainage network intact to allow for the free movement of the aquatic species. • Perform earthworks at nearby water bodies in dry periods.
Improper solid waste disposal	• Dump disposals in an approved location to avoid contact with terrestrial and aquatic habitats. • Strictly adhere to the environmental management plan. • Introduce a point system to encourage the use of recycled material. • Educate project stakeholders with reduce, reuse, and recycle circular built environment principles in highway construction.
Noise and vibration	• Update the crack survey and maintain the records. • Do the pre-crack survey before the commencement of construction works. • Involve with joint inspections, and monitoring mechanisms with government regulatory bodies such as the road development authority. • Contractors can transfer the losses caused by environmental risk occurrences through insurance companies.

construction enterprises. In the late 1990s, the *Lean Construction Institute* was established in Denmark, which regularly holds seminars on the lean construction theory. Since then, more scholars, institutions, and construction companies worldwide have extensively developed lean construction.

Lean construction applies lean thinking to the production process of construction projects, significantly transforming the manufacturing industry's production methods and management concepts. In lean construction, the construction process is viewed as a unique production process with many discontinuous tasks, and each task adds value to the construction product. However, many of these tasks can also lead to substantial waste. By maximizing the efficiency of each task, lean construction reduces construction time, minimizes changes and claims, lowers project costs, and ultimately benefits sustainable construction.

In 1996, Womack and Jones summarized five basic principles in lean thinking: the first is to determine the value correctly; the second is to identify the value stream; the third is the flow of the value stream; the fourth is demand-driven production; and the fifth is to pursue perfection. When applying to the lean construction theory, these five basic principles are formed to correctly determine the value of the project, to identify the value stream of the construction project, the flow of the value stream of the construction project, the pull of customer demand of construction projects, and to continuously improve construction projects towards perfection. Therefore, lean construction is a construction management model in the construction industry, guided by lean thinking, using a variety of lean management tools to transform the construction process, eliminate waste in the value stream of construction products, improve product quality, enhance the flexibility of construction management, and provide products that meet the owner's requirements, which can prompt traditional enterprises in the construction industry to transform into lean organizations and improve the competitiveness of the enterprises.

Previous studies have investigated many benefits of implementing lean construction, for example, Bajjou and Chafi (2018) and Ahmed et al. (2021). The primary benefits can be summarized in the following:

- Better project quality
- Improving safety
- Improving the environmental performance
- Reducing overall project duration
- Reducing construction costs and faster turnover of projects
- Enhancing the sustainable development idea in the project
- Increasing productivity and customer satisfaction

3.5.2 Ways to implement lean construction

Implementing lean construction involves a series of procedures that aim to maximize value while minimizing cost and waste during the construction process. Previous studies and practices have identified many techniques that can be applied in promoting lean construction. The main techniques are summarized and basically described in Table 3.4. Some techniques are considered management methods that can be used for solving various management issues and improving construction performance, such as Pareto analysis, Plan-Do-Check-Act (PDCA), Kaizen (continuous improvement), and 5 whys. Some techniques are mainly for improving specific management aspects, such as Six Sigma and Failure Mode and Effects Analysis for quality management, and JIT (just-in-time) and concurrent engineering for schedule management. Few techniques can simultaneously save construction materials and reduce construction waste to a certain extent, such as 5S and work standardization.

3.5.3 From lean construction to sustainable construction

For both sustainable construction and lean Construction, the first important target is to achieve construction project objectives, such as project cost, schedule, and quality objectives. However, they focus on different aspects and pathways during the construction process. Sustainable construction is more focused on environmental and social considerations, while lean construction has a broader scope in improving construction efficiency, effectiveness, and environmental performance. Moreover, sustainable construction sometimes involves upfront investments for long-term benefits, particularly in choosing green materials from LCA, while lean construction aims for immediate efficiency gains and adding value. Despite their differences, they are not mutually exclusive, and there can be synergies when implementing both principles to create construction projects that are both environmentally responsible and highly efficient.

Even though lean construction concentrates on improving the overall construction process, it also considers saving materials and reducing waste as key principles. Therefore, lean construction could also promote the implementation of sustainable construction, such as through the following ways.

- Lean construction aims to reduce costs, such as saving construction energy and resource consumption, which will reduce the impacts on the environment.
- Lean construction targets to eliminate waste, which is in alignment with the key objectives of sustainable construction.
- Lean construction optimizes the construction process by enhancing construction efficiency and reducing duplication and ineffective and low-effective work. It will streamline the construction process, reduce delays, and minimize resource usage.
- JIT principles are key principles in lean construction to minimize excess inventory. It can avoid unnecessary storage and promote efficient resource utilization.

Table 3.4 Primary lean construction techniques

Lean construction techniques	*Brief description*
5s	Stands for sort (*Seiri*), set in order (*Seiton*), shine (*Seiso*), standardize (*Seiketsu*), and sustain (*Shitsuke*). Using visual controls is an effective process for waste removal from construction sites.
JIT	It is a technique to reduce the time flow of production, particularly for the response time from suppliers to the end-users. It is also a method of thinking, working, and controlling waste in construction.
Concurrent engineering	It is a method of performing various works parallel to multi-disciplinary teams aiming for optimization of the engineering cycle mainly for efficiency.
PDCA	It is an interactive process of improving or managing new and existing construction issues. It means plan (set up a plan and expect results), do (execute the plan), check (verify anticipated result achieved), and act (evaluate; do it again).
Six sigma	It is a set of techniques for mainly improving quality by eliminating defects and minimizing variations in the construction process.
Pareto analysis	It is a bar chart method of analysing data by the frequency of causes or problems of any operational process. It visually describes that what is the importance or best level of any process or situation.
LPS (last planner system)	LPS is a collaborative planning process that involves work trades in planning in greater and greater detail as the time for the work to be done gets closer. In construction, LPS is considered an effective tool to control workflow and reduce project variability.
Kaizen (continuous improvement)	This is a Japanese production and process philosophy of continuous improvement. It allows continuous analysis and improvement of the process in terms of resources, quality, supply-demand, team performance, and so on.
Kanban (pull system)	This Japanese word stands for a sign or card. It is used in controlling the amount of material/components in the stock. It regulates the movements or flow of resources so that parts and supplies are ordered and released as they are needed.
FMEA	This is a step-by-step process for determining potential failure in construction. These failures are also ranked to prioritize their level of effects and consequences to take actions to eliminate them, starting with the highest-ranked ones.

(*Continued*)

Table 3.4 (Continued)

Lean construction techniques	*Brief description*
FIFO line (first in, first out)	This is an approach for handling work requests in order of flow from first to the last.
Value stream mapping	It is a method that allows analysing, documenting, and improving any flow of process by visualization with a great improvement opportunity.
Poka-yoke (error-proofing)	This is a Japanese method, mainly for quality management. It is a mechatronics technique to detect errors and defects with the aim of zero defects in the process.
First-run studies	It is a tool for the trial execution of a process with a specific final goal to decide the best means, strategies, and sequences, among others to perform it.
Visual management	It is an information communication method that enhances the efficiency and clarity of the process through visual signals.
TQM (Total quality management)	It is a method to achieve the organization's targets and customer requirements by integrating all the organizational functions.
5 whys	It is a problem-solving technique used to identify the root causes of a targeted problem. The questions are usually specific to the project and are not limited to five questions.
Work standardization	It is a manufacturing documented procedure that captures best practices. This "living" documentation is undemanding to modify, mainly aiming to enhance construction efficiency and save waste. It is one of the 5s contents.
Fail-safe for quality	It is a method that provides alerts for potential defects or risks in construction. It is similar to the poka-yoke technique but it is extended to safety management.
Team preparation	This is a process of providing training to the employees on waste, TQM flow, and standardizing work.
Daily huddle meetings	This is a technique of the everyday meeting process of the project team to accomplish workers' involvement with project awareness and problem-solving contribution.

Source: Bajjou and Chafi (2018) and Ahmed et al. (2021)

- The management ways of lean construction, such as continuous improvement, PDCA, and 5S, can contribute to sustainable construction management. For example, PDCA in sustainable construction can help develop sustainable construction plans, implement, check, and then promote them in a cyclic way.
- Lean construction attempts to standardize construction and industrial construction, which can significantly save construction materials and energy and then reduce waste and emissions.

Ogunbiyi et al. (2014) explored the contribution of lean construction to sustainable construction through a survey. The results investigated the primary benefits and implementation techniques associated with synchronizing lean and sustainable construction, which are shown in Table 3.5. The key linked areas between lean and sustainable construction include waste reduction, environmental management, value maximization, and health and safety improvement. In summary, even though it should not be over-emphasized that lean construction techniques benefit sustainable development in the construction industry, it can be evidently predicted that implementing lean construction could significantly promote sustainable construction practices and performance. Here, the author selects three techniques, including 5S, JIT, and concurrent engineering.

Table 3.5 The ranked benefits and techniques of implementing lean and sustainable construction.

Rank	*Ranked benefits of synchronizing lean and sustainable construction*	*Ranked techniques of lean construction for establishing sustainability*
1	Improved corporate image	JIT
2	Increased productivity	Visualization tool
3	Waste reduction	Daily huddle meetings
4	Energy reduction	Value analysis
5	Improvement in sustainable innovation	Value stream mapping
6	Improved process flow	Total quality management
7	Material reduction	Fail-safe for quality
8	Reduced cost and lead time	5S
9	Improvement in health and safety	Total prevention maintenance
10	Improvement in environmental quality	First-run studies
11	Water reduction	Last planner
12	Increased sustainable competitive advantage	Concurrent engineering
13	Increased compliance with customers' expectation	Pull approach
14	Increased employee morale and commitment	Kanban
15	-	Kaizen
16	-	Six Sigma

Source: Ogunbiyi et al. (2014)

5S

In lean production theory, the simplest and most effective method to improve the level of on-site management is the 5S management method: namely Sort (*Seiri*), Set in Order (*Seiton*), Shine (*Seiso*), Standardize (*Seiketsu*), and Sustain (*Shitsuke*). The 5S is a management methodology for improving operational efficiency and sustainability in construction companies. The technique aims to reduce waste, decrease material and equipment operation time, enhance productivity, and improve customer satisfaction. In practical applications, the use of 5S can benefit several different aspects of a construction company's long-term sustainability, such as:

- Clearly marking and designating areas, and identifying item locations to prevent loss or waste
- Integrating standardized company processes and executing them on-site
- Effectively developing employee skills and continuous improvement
- Helping to minimize tedious work and enhancing employee morale and creativity

Nguyen and Ogunlana (2006) explore the use of 5S to improve construction site performance, particularly in enhancing productivity, reducing waste, and improving safety on five construction projects in Singapore to identify how 5S can benefit the construction industry. Through implementing 5S practices, the project team was able to improve their overall site performance by reducing delivery and operation time for materials, tools, and equipment, maintaining proper storage of materials, and eliminating unnecessary items from the working area.

Just-in-Time (JIT)

JIT production is an important concept of lean management that emphasizes the efficient use of resources by delivering materials and components at the precise time the construction needs. JIT is based on production equalization during the whole supply chain. It aims to minimize excess inventory and reduce the need for large storage areas, as far as possible to achieve high quality, low cost, low resource consumption, and the shortest production and transportation delivery time. Accordingly, it can promote the achievement of sustainable construction practices. Following are some to achieve sustainable construction by using the JIT approach.

- Reduced waste and energy consumption: In sustainable construction, JIT can target to physically stock zero inventory on construction sites and reduce the large storage area and second transport. JIT directly translates to a reduction in waste, as materials are ordered and used in quantities that closely match project requirements. By delivering materials exactly when

they are needed, transportation and handling are minimized. This not only reduces the carbon footprint associated with transportation but also saves energy by reducing the need for storage facilities. Consequently, this can result in less material waste and a more sustainable use of resources.
- Optimized resource and equipment utilization: JIT principles can be applied to optimize human resources and equipment. By scheduling construction activities to align with the availability of skilled workers and specialized equipment, overall productivity can be improved. This ensures that resources are used efficiently, minimizing downtime and idle periods. Efficient logistics contribute to overall energy savings and make the construction process more sustainable.
- Improved collaboration and commitment to sustainable construction. The JIT implementation embodies the capabilities and provides the foundation for sustainable construction. JIT promotes better communication and collaboration among project stakeholders, including suppliers, contractors, and builders. This can result in more streamlined processes, reduced delays, better sources of sustainable materials, and increased overall efficiency. The improved coordination can lead to sustainable construction practices by minimizing unnecessary resource consumption. Project stakeholders will become more integrated into the construction process and share the commitment to sustainable practices.

Concurrent engineering

Concurrent engineering is a systematic mode of designing products and processes (including manufacturing processes and supporting processes) in a parallel and integrated manner, which allows developers to consider all factors in the product life cycle from the very beginning, including quality, cost, schedule, and user needs. It is common to integrate a concurrent engineering approach in lean construction, as it can speed up the construction processes, enhance collaboration, and achieve greater efficiency in completing projects. The four key elements of concurrent engineering are as follows:

- Involving all construction and service teams and other stakeholders of the entire project life cycle early in the design phase to reduce iterations and ensure that various targets and constraints are considered simultaneously.
- Implement simultaneous construction by allowing different subcontractors to work on-site concurrently, reducing waiting times, and enhancing efficiency.
- Cross-train employees to enhance flexibility and adaptability within the organization, aiming to enable them to contribute effectively to concurrent engineering activities, promoting a more versatile and collaborative work environment.
- Foster a culture of continuous learning and improvement within the project, encouraging teams to share insights, feedback, and lessons learned for ongoing enhancement of processes.

3.5.4 *Case study: the lean construction platform leading the digital transformation and upgrading of the entire construction chain*

Decai Group Co., Ltd. in China was established in 1999. It is one of the largest companies in the national architectural decoration industry. Since its establishment, Decai Group has faced prolonged challenges such as a highly competitive and risky market, poor compliance with industry standards, complex management characteristics, weak project control, delayed delivery chains, and low profit margins. To address these challenges and foster company development, Decai Group established a "Lean Construction Platform". This platform allows main contractors, material suppliers, subcontractors, and engineering consultants to collaborate in real time on a single platform. It facilitates collaboration, quality control, and safety risk management across multiple projects, departments, and branches in different geographical areas, driving a transformation in production modes. The platform engages all participants and professions, enables precise control of project costs, and promotes collaborative construction throughout the entire construction process. By using this platform, Decai Group can focus on top-level management, operations, and business performance, with the following objectives:

Unified management and technical empowerment

By establishing a unified background operation, data management, and technical empowerment platform, lean management facilitates the continuous optimization of management processes and effective collaboration. This comprehensively improves functional management levels and enhances management efficiency.

Front-end business operation system

The platform offers an excellent front-end business operation system that enables business demand analysis, uncovers business potential, prevents and controls business risks, and helps the company establish a data-driven and agile operational framework that continuously improves performance.

Public service platform for the entire industry chain

The platform serves as a public service platform for the entire company industrial chain, supporting the full life cycle of decoration projects. It encompasses more than 3,000 enterprises in industries such as design, construction, subcontracting, and material supply. It enables participating enterprises to efficiently and effectively complete project design, procurement, construction, usage, and operation tasks, facilitated by online procurement and bidding, cloud-based building information modelling (BIM), and "one-click" diagram and order generation.

Innovative business models

The platform explores and experiments with innovative business models, starting with investment and cooperation. It provides digital business opportunities for the company to apply advanced digital technologies such as big data and digital twins, and to explore new models like industrial construction. Consequently, the company has built a new industrial ecosystem focused on resource sharing and achieving multi-party win-win scenarios.

For more information, visit the website of Decai Group (http://en.decaigroup.com/touzizheguanxi/).

3.6 Integrated project delivery in sustainable construction

3.6.1 What is the integrated project delivery method?

The Integrated Project Delivery (IPD) definition from the American Institute of Architects (AIA) is widely accepted as "a project delivery approach that integrates people, systems, business structures, and practices into a process that collaboratively harnesses the talents and insights of all participants to optimize project results, increase value to the owner, reduce waste, and maximize efficiency through all phases of design, fabrication, and construction". This integrated approach enables all participants to optimize and enhance the project's value, minimize waste, and maximize overall efficiency throughout the project life cycle of design, construction, and operation.

According to the IPD guide of AIA, it is grounded in the principles of "mutual respect and trust, mutual benefit and reward, collaborative innovation and decision making, early involvement of key participants, early goal definition, intensified planning, open communication, appropriate technology, and organization and leadership". The key framework for implementing IPD is illustrated in Figure 3.6. Therefore, when comparing IPD to traditional delivery models, the most significant change is not only in the project delivery and management approach but, more importantly, in altering the behaviours of the project team members.

IPD is an innovative approach that involves close collaboration among key project stakeholders such as owners, designers, contractors, and suppliers. This collaboration enhances work efficiency across different project management teams. The main characteristics of IPD include the following:

- **Multi-party relationship contracts for IPD:** These contracts ensure that all parties jointly assume responsibility for project costs, schedule, and quality, forming a collaborative team that shares information, risks, and benefits. The focus of these contracts is on the implementation process rather than the final product. IPD contracts should detail the rights and obligations of each party during project implementation to ensure smooth project execution.

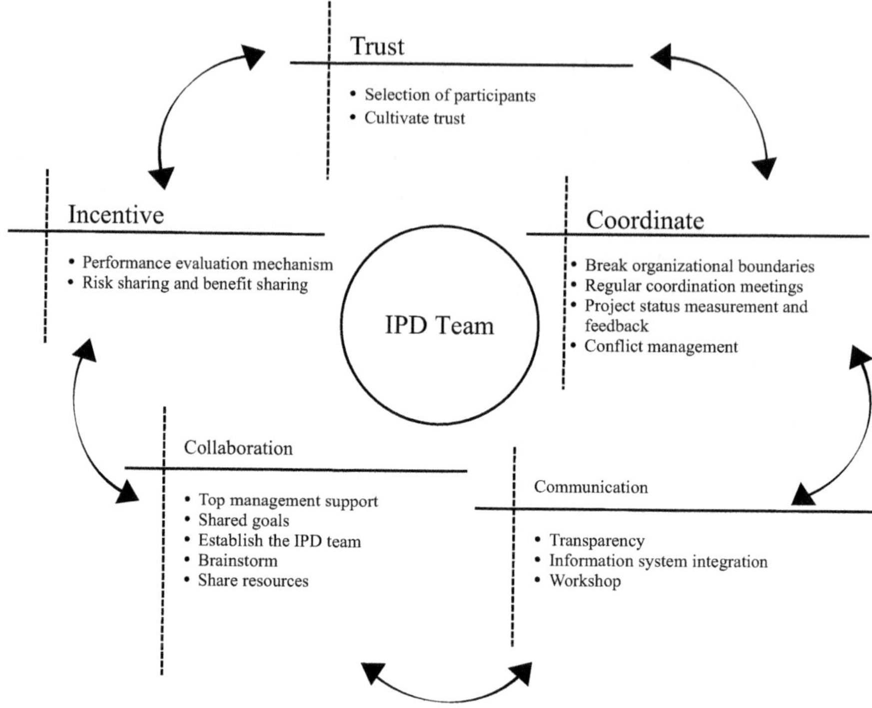

Figure 3.6 The implementation frameworks of IPD.
Source: Yang and Du, 2015

- **Collaborative team relationships:** The IPD model emphasizes collaborative relationships among various parties throughout the project life cycle. This approach incentivizes each party to serve the overall project interests actively, leading to optimal outcomes. All parties must achieve the overall project goals for their individual returns to be maximized, improving overall project performance. This benefit distribution model encourages parties to work closely together and share information, aligning their interests to reduce project risks and costs. Sharing information and experiences fosters a learning-oriented team, enhancing trust, communication, collaboration efficiency, and true team cooperation.
- **Management and technical means:** IPD construction projects involve frequent information sharing, close collaboration, and strict quality control regarding costs, schedules, and project completion. The use of advanced management methods such as lean construction management and BIM ensures the smooth implementation of projects.
- **Reducing project delivery risks:** Unlike traditional construction models, where participants often shift risks among themselves driven by self-interest, IPD brings together owners, designers, contractors, and other project participants to create a cooperative system based on shared values and mutual

trust. This collaborative system encourages all participants to manage risks, share risks, and share rewards collectively, leading to a win-win outcome. In an IPD project, risks are equitably distributed among participants, integrating risks and uncertainties with the overall project objectives to promote fairer risk management. The IPD model offers incentives and penalties to project participants, aligning individual objectives with the overall goal of project success and reducing project risks.

3.6.2 Integrated project delivery for sustainable construction

IPD aims to remove obstacles to collaboration, align the interests of all stakeholders, and incentivize actions that add value to the project. By prioritizing the project's overall success as a shared goal, IPD meticulously drafts contracts to protect the legal interests of all involved parties and achieve the best possible project delivery outcomes. This approach involves all parties signing relational contracts and forming highly collaborative project teams. Unified interests enable cross-organizational and cross-disciplinary team members to work, communicate, and learn together daily in the same workspace. Open, sincere, and transparent communication helps project teams establish trust and respect, fostering a sense of teamwork and significantly reducing information loss during communication. When one party encounters challenges, other parties can quickly assemble to offer strong support. Therefore, IPD can promote sustainable construction practices in the following ways:

- Developing the goals of sustainable construction: Before starting an IPD construction project, the owner, designers, general contractors, subcontractors, and suppliers collaboratively negotiate and establish project objectives, including time, quality, cost, and sustainable construction goals. During contract negotiation and goal-setting phases, IPD participants ensure that the goals fully leverage each participant's skills and knowledge, aligning with their interests. This improves overall project efficiency, reduces waste, and achieves optimal project sustainability outcomes.
- Planning and designing sustainable construction strategies at the early stage: IPD encourages all stakeholders to participate in planning and workflow development during the project's initial stages. This facilitates the early identification of potential challenges and opportunities for sustainable construction, enabling better integration of sustainable design strategies.
- Sharing responsibilities, risks, and rewards of sustainable construction: IPD's contractual framework binds the responsibilities, risks, and rewards of project participants together, fostering a team-oriented approach that incentivizes collaboration and promotes integration in sustainable construction. This innovative project delivery system focuses on the collective efforts of all participants, ensuring they work together towards the sustainability of project construction activities.
- Establishing a constructive collaborative environment for sustainable construction: IPD promotes cross-disciplinary communication and

collaboration, reducing misunderstandings and conflicts, and enhancing the team's consensus on sustainable construction. With all major stakeholders involved early in the project, they collectively assume responsibility for its success in sustainable construction. This encourages the team to seek sustainable solutions throughout the design, construction, and operational phases. Establishing a constructive collaborative environment requires long-term cooperation, sustainable business growth, a high degree of trust among participants, and a shared willingness to assume risks.
- Producing more resources and profits for sustainable construction: Traditional project delivery models derive profits from the difference between the contract price and the actual cost, incentivizing cost reductions that may lead to inefficiency and delays. In contrast, the IPD model aligns the rewards of all participants with the overall achievement of project objectives, linking profit to project performance. Profits are distributed among participants according to predetermined proportions from the total revenue generated during the design and construction process, with additional rewards allocated based on the extent to which project requirements are met by the owner. Through contracting sustainable construction objectives and collaboration, IPD helps identify unnecessary costs and waste in the project, providing more resources and profit for sustainable construction.
- Using advanced sustainable construction technologies and approaches: The IPD approach encourages the use of new technologies and innovative methods to enhance project sustainability. For example, advanced BIM technology can optimize design and construction processes. Successful application of advanced sustainable construction technologies and approaches requires the support of all project participants. In traditional models, owners, designers, and contractors operate independently, which results in fragmented management. BIM technology is often confined to a single party or stage. The IPD model integrates all parties to maximize the benefits of advanced technologies within one system, serving as an essential pathway to achieving high-performance construction.

In summary, IPD facilitates more sustainable project design and construction by promoting cross-team collaboration, shared responsibility, and goal-oriented approaches. This model allows project teams to better address the challenges of sustainable construction while improving project quality and efficiency. The IPD model represents a trend towards modern project delivery methods and has demonstrated excellent efficacy in sustainable project applications.

3.6.3 *Case study*

IPD practice in Shanghai Tower, China

Shanghai Tower, located in Shanghai, is the tallest building in China and the second tallest in the world as of 2019, standing at 632 metres. The construction of the project began in November 2008, with civil engineering work completed by the end of 2014, and the building was put into trial operation in January 2017. The project encompasses a total floor area of 575,000 square metres, with a main structure height of 580 metres, and consists of 121 floors above ground and 5 floors underground. The total investment amounted to approximately 148 billion yuan. This large and complex ultra-high-rise project features nine functional areas, seven structural systems, and more than 30 mechanical and intelligent subsystems. It employed over 30 consulting companies for architectural structure, curtain wall, mechanical, and fire protection, along with numerous professional subcontracting teams. The involvement of different experts and a variety of workforces necessitated an effective coordination mechanism.

To address this need, Shanghai Tower introduced a comprehensive IPD management approach. IPD, a contemporary project delivery approach, emphasizes collaboration, innovation, and efficiency. In the Shanghai Tower project, the IPD management platform was established utilizing digital technologies such as BIM and big data. Led by the owner and integrated with participating teams, this approach and platform allowed organizations to optimize their resources, reduce waste, create more value for clients, and address the "information island" phenomenon caused by information asymmetry in traditional project management. Through this management platform, the participating parties formed a circular structure for collaborative work, ensuring real-time information exchange and seamless management between organizations. This approach facilitated the non-lossy transmission and full sharing of project information throughout the planning, implementation, operation, and other project life cycles.

BIM-based IPD for sustainable procurement in Sri Lanka

Rosayuru et al. (2022) investigated and verified the suitability and benefits of the BIM-based IPD procurement system for sustainable procurement in the construction industry. The system, which combines sustainable procurement and BIM-based IPD, can bring significant technical, managerial, environmental, economic, and social benefits to sustainable construction. The ranking of the BIM-based IPD principles to achieve sustainable procurement is shown in Table 3.6.

3.7 Conclusion

The significance of sustainable construction has become increasingly recognized by professionals and organizations. Consequently, various sustainable construction project management methods have been developed to facilitate its implementation. This chapter outlines several critical management methods that have been or can be utilized to manage sustainable construction effectively.

Table 3.6 Ranking the contribution of BIM-based IPD principles to achieve sustainable procurement (Rosayuru et al., 2022).

BIM-based IPD principles	*Rank*
Mutual trust among the parties and alignment of objectives of all	1
Early involvement of key team members	2
Open communication and effective co-ordination among parties	3
Positive attitude of the project participants (including collaboration and flexibility)	4
Professional ethics	5
Early consideration of logistics and other buildability issues	6
Awareness and assessment of risks and rewards	7
Appropriate technology (BIM, cloud servers, teleconference tools, automated production)	8
Clarity of management decisions	9
Safety plan, project labour agreement, alternative disputes resolution for workers' compensation should be part of the early	10
Agreed, clear, and quick process for dispute resolution	11
Agreed mechanism for performance appraisal and pay outs	12
Intensified planning	13
System thinking and lean thinking	14
Whole life value assessments that include organizational outcomes	15
Collective responsibility instead of personal responsibility	16
Legal implications and consistency with public law requirements	17
Compatible organizational cultures	18
Frequent formal and informal meetings for continual improvement	19
Project-specific insurance	20
Pioneering role of the owner/client	21
Language directing the parties to collaborate upon developing project	22
Separate project bank account for each project	23

One such method is LCM, which considers the entire life cycle of a project, from conception through to demolition or reuse. By examining the entire life cycle, it is possible to identify areas for improvement and develop comprehensive strategies for sustainable construction.

Sustainable procurement and supply involve selecting suppliers and contractors committed to sustainable practices. This means choosing companies that use sustainable materials and have a demonstrated commitment to reducing waste and emissions. This approach integrates environmental, social, and economic considerations into the procurement process, ensuring that environmental, social, economic, and project requirements are met.

Risk management for sustainable construction is crucial for the successful implementation of sustainable projects. It promotes the integration of sustainable practices, minimizes negative environmental impacts, and enhances overall project efficiency. Continuous improvement in risk management is essential to adapt to evolving risks and emerging trends, ensuring that projects remain aligned with sustainable objectives and regulatory requirements.

Lean construction is another significant approach to sustainable construction. This methodology applies principles from lean manufacturing to make the construction process more efficient. It can lead to significant improvements in corporate image and efficiency, waste and energy reduction, and overall sustainability. Various lean construction techniques, such as 5S, JIT, and concurrent engineering, offer multiple pathways to promote sustainability across different construction activities.

IPD is a collaborative project delivery method that brings together all project stakeholders early in the construction process to work towards shared sustainable construction goals. IPD fosters communication, teamwork, and innovation among architects, engineers, contractors, and owners. By promoting transparency, accountability, and shared responsibility, IPD can drive the successful implementation of sustainable construction projects while maximizing value for all stakeholders.

Importantly, these methods can be comprehensively and simultaneously applied to a single project. IPD can be used to construct and deliver projects, while LCM ensures that every phase of the project is underpinned by sustainable principles. Sustainable procurement and supply methods can identify and contract participants who adhere to sustainable practices throughout the project. Integrating risk management into these practices allows stakeholders to proactively identify, assess, and mitigate potential risks, safeguarding the project's sustainability objectives. Combined with lean construction principles and IPD, these strategies streamline the construction process to eliminate waste and enhance productivity, fostering a collaborative environment that nurtures innovation and shared vision among all project participants. In summary, these holistic approaches to sustainable construction ensure that projects are efficient, resilient, and environmentally responsible, collectively contributing to a more sustainable built environment.

References

Abhayantha, K. I. L., Perera, B. A. K. S., Perera, H. A. H. P., & Palliyaguru, R. S. (2023). Management of environmental risks in highway construction projects in Sri Lanka. *Construction Innovation*. https://doi.org/10.1108/CI-08-2022-0202.

Ahmed, S., Hossain, M. M., & Haq, I. (2021). Implementation of Lean construction in the construction industry in Bangladesh: Awareness, benefits and challenges. *International Journal of Building Pathology and Adaptation*, *39*(2), 368–406.

Bajjou, M. S., & Chafi, A. (2018). Lean construction implementation in the Moroccan construction industry: Awareness, benefits and barriers. *Journal of Engineering, Design and Technology*, *16*(4), 533–556.

Commonwealth of Australia. (2021). *Sustainable procurement guide—A practical guide for Commonwealth entities.* https://www.dcceew.gov.au/sites/default/files/documents/sustainable-procurement-guide.pdf

DEFRA. (2006). *Procuring the future. Department for environment, food and rural affairs*. London. https://assets.publishing.service.gov.uk/media/5a78e19040f0b6324769ae5f/pb11710-procuring-the-future-060607.pdf

El-Sayegh, S. M., Manjikian, S., Ibrahim, A., Abouelyousr, A., & Jabbour, R. (2021). Risk identification and assessment in sustainable construction projects in the UAE. *International Journal of Construction Management*, *21*(4), 327–336.

Ershadi, M., Jefferies, M., Davis, P., & Mojtahedi, M. (2021). Achieving sustainable procurement in construction projects: The pivotal role of a project management office. *Construction Economics and Building*, *21*(1), 45–64.

Nguyen, T., & Ogunlana, S. (2006). Improving construction site performance using 5-S: An ardent survey—case study. *Journal of Construction Research*, *7*(2), 295–305.

Ogunbiyi, O., Goulding, J. S., & Oladapo, A. (2014). An empirical study of the impact of Lean construction techniques on sustainable construction in the UK. *Construction Innovation*, *14*(1), 88–107.

Rosayuru, H. D. R. R., Waidyasekara, K. G. A. S., & Wijewickrama, M. K. C. S. (2022). Sustainable BIM based integrated project delivery system for construction industry in Sri Lanka. *International Journal of Construction Management*, *22*(5), 769–783.

Womack, J. P., & Jones, D. T. (1996). *Lean thinking: Banish waste and create wealth in your corporation*. Simon & Schuster US.

Yang, Y., & Du, J. (2015). IPD Mode and Its Management Framework in Project Management (In Chinese). *Journal of Engineering Management*, *29*(1), 107–112.

4 Smart construction technologies towards sustainable development

4.1 Introduction

With the continuous development of advanced technologies, researchers and practitioners are increasingly interested in leveraging these innovations for industry development and environmental protection. These technologies not only enhance productivity and efficiency but are also crucial for fostering sustainability within the construction sector.

The construction industry is particularly embracing digital transformation and integrating innovative technologies to achieve more efficient, cost-effective, and environmentally friendly methods, promoting sustainable development. Sustainable construction technologies can integrate environmental protection, energy conservation, and efficiency enhancement in building construction. They have the ability to reduce environmental impact, improve economic performance, decrease resource waste, increase work efficiency, and foster green construction and management.

This chapter will primarily discuss sustainable construction techniques, focusing on 3D printing in construction, BIM-based sustainable construction, industrialized construction, prefabricated construction, and other advanced technologies such as machine learning, digital twins, big data, blockchain, and the Internet of Things (IoT). It will showcase the innovation, potential, and importance of advanced construction technologies in facilitating sustainable construction practices and ensuring a sustainable future for the construction industry in Asia.

4.2 3D Printing construction

4.2.1 What is 3D printing construction?

3D printing construction technology combines 3D printing robots with cement, concrete, and other construction materials, offering a novel approach to the construction sector. This rapid prototyping technology uses digital models and computer graphics processing to convert 3D building models into tangible structures layer by layer. The principle of 3D printing construction

DOI: 10.4324/9781003202660-4

involves extruding configured concrete and other slurry from the nozzle of the printing equipment to create designed components or structures. This technology has evolved from printing small elements to directly printing entire buildings, and from simple components to complex structures, high-rise buildings, and bridges. Printed products demonstrate improved strength, seismic resistance, and waterproofing properties. Unlike traditional concrete forming processes, 3D printing construction is one of the latest and fastest mould-free forming technologies, eliminating the need for template support. Importantly, this technology promotes sustainable construction by reducing project duration, enhancing building precision, and minimizing material waste. In summary, as a new manufacturing technology, 3D printing is gaining importance as a viable and environmentally friendly solution for sustainable construction, despite the high initial cost of equipment and limited suitable materials.

Three-dimensional printing materials must meet several requirements. First, they must be printable, allowing for accurate layer-by-layer construction. Second, they should have sufficient strength, durability, flexibility, elasticity, and compatibility for building structures. Third, cost-effectiveness is essential to balance material performance with budget considerations. Materials may also need to meet special requirements such as colour, appearance, and environmental resistance. In line with sustainable development goals, materials should have low embodied energy and a low carbon footprint. Sustainable materials with low carbon footprints, such as fly ash, geopolymer materials, and recycled glass, are already used in the industry. Commonly used 3D printing materials include concrete, metal, and clay.

Three-dimensional printing with concrete

Unlike traditional concrete moulding, 3D-printed concrete is based on 3D data and eliminates the intricate moulding process of concrete construction. The concrete material is extruded through a concrete printing nozzle and assembled layer by layer, to form the concrete printing model. The ability to 3D-print concrete structures offers a wide range of benefits in terms of sustainability, especially in terms of saving labour costs and environmental friendliness.

Three-dimensional printing with clay

Clay is a natural material that comes from the earth. The clay used in 3D printing is typically a mixture of clay minerals, water, and other additives. The composition may vary depending on the specific requirements of the 3D printing process. Three-dimensional printing with clay involves using a specially formulated clay material that can be extruded layer by layer to create 3D objects. The objects are constructed through a clay printing process using the traditional strip clay extrusion in a 3D printer. Therefore, transforming traditional manual clay extrusion into robot movements can perform these movements with more precision, agility, and consistency.

Three-dimensional printing with metal

Three-dimensional printing with metal offers much greater design flexibility than traditional metal manufacturing processes. Complex shapes and curves can be created, even if they are difficult or even impossible to make with traditional procedures. Moreover, using 3D printing in metal components can reduce the number of different parts and assemblies, and generally requires fewer raw materials than that required in traditional techniques, making it more cost-effective and eco-friendly. More importantly, with the ability to print metal parts using recycled metal powders, 3D printing can reduce the environmental impact of mining and the production of new metal parts. Accordingly, it will help to reduce the amount of metal waste produced in construction activities and promote the circular economy.

Case: China's first three-dimensionally printed retractable bridge unveiled in Shanghai

In 2021, China's first 3D-printed retractable bridge was unveiled in Shanghai's Wisdom Bay Innovation Park (see Figure 4.1). This Bluetooth-controlled bridge measures 9 metres in length, 5 metres in width, 1.1 metre in height, and weighs 850 kilograms. The successful creation of this bridge is considered a step forward in the development of 3D printing technology for retractable architecture.

The bridge is composed of 36 triangle panels, each with a unique design that resembles outward ripples. Produced over a three-day period, these panels are made from an environmentally friendly composite carbonate polyester material. The bridge is segmented into nine sections, and its retractable feature is operable via Bluetooth, enabling the structure to unfold in a spiral shape over the water in under one minute. Moreover, the bridge is equipped with a gravity-sensitive automatic warning system to prevent overloading.

4.2.2 Three-Dimensional Printing technology for sustainable construction

Currently, the construction industry, as a labour-intensive industry, faces huge challenges due to a labour shortage worldwide. The emergence of 3D printing construction technology undoubtedly promotes the sustainable development of the construction industry. For example, 3D printing construction eliminates the tedious procedure of mould support and mould removal in concrete construction, greatly simplifying the construction process. Moreover, it can fully utilize material, reduce material usage, ensure construction quality, and decrease the generation of construction waste and the repetition of construction processes. Additionally, 3D printing machinery requires fewer labourers

Figure 4.1 China's first 3D-printed retractable bridge.
Source: 3dprintingindustry

than traditional construction methods. The applications and advantages of 3D printing construction are summarized in this chapter as follows.

Environmentally friendly construction

Compared to traditional construction techniques, 3D printing construction technology is more environmentally friendly. Apart from the printing material itself, the technology eliminates the need to transport various other materials to construction sites. For instance, the printing material is typically a reformed cementitious mixture shaped into structures, which would greatly improve the work environment by avoiding dusty and noisy conditions and reducing the need for transporting concrete and formwork.

Moreover, 3D printing technology can improve the sustainability of the construction process by reducing waste. As a precise and accurate digital printing process, it effectively consumes the necessary materials during construction, avoiding unnecessary waste of resources and the waste of cutting materials for size differences. In addition, the materials required for 3D printing can be recycled from construction waste, where separated waste can be combined with high-strength concrete in a certain proportion to form the materials for 3D printing construction. Therefore, this can greatly reduce building waste and carbon emissions compared to traditional construction methods. The application of 3D printing technology is conducive to an environmentally friendly construction technology.

Saving construction time

Three-dimensional printing technology has emerged as a promising solution to the housing demand driven by rapid social development. It has the potential to revolutionize architecture and streamline the construction supply chain, leading to significant time and cost savings. For example, in reinforced concrete structures, building moulds typically consume 50%–75% of total construction time. Three-dimensional printing technology eliminates the need for building moulds, saving considerable manpower and time. The entire printing process requires only a 3D printer, a computer, a few labourers, and the necessary printing materials. Consequently, stakeholders across the construction industry, including suppliers, distributors, construction teams, and architects, can benefit from the advantages offered by 3D printing technology.

Reducing construction labour and safety accidents

Three-dimensional construction presents the potential for autonomous processes, as computer-controlled printing machines significantly reduce labour requirements compared to traditional construction methods. For example, a 3D printer can execute complex construction projects with minimal skills required, unlike traditional methods that rely on a range of specialized and skilled labourers. This not only streamlines the construction process but also enhances safety on construction sites. With fewer labourers and reduced site complexity, the risk of safety accidents such as falls, collapses, or being struck by objects is significantly mitigated. Consequently, adopting 3D printing construction technology can contribute to creating safer construction sites and improving overall work comfort for construction workers.

Innovative construction for special requirements

Architects often conceive ideas that surpass the capabilities of traditional construction methods, but 3D printing technology bridges this gap by turning these innovative designs into tangible realities. Regardless of the complexity of the structural design, as long as the blueprints are available, 3D printing can address challenges that traditional methods cannot solve. This not only fosters flexibility and innovation in building design but also enables more individuals to acquire personalized homes that align with their specific needs, thanks to the ability to create concrete structures in 3D. Moreover, integrating building service systems like heating, insulation, water supply, and electricity into the 3D printing process enhances efficiency and reduces waste. Hollow wall printing minimizes material usage while improving insulation, and the utilization of 3D-printed canals for water transport further optimizes resource utilization. This ultimately leads to more sustainable and intelligent buildings without the need for time-consuming on-site installations. In summary, the versatility of 3D printing has facilitated the creation of intricate

Figure 4.2 India's first 3D-printed house.
Source: Interior company

and unconventional building structures. Therefore, 3D printing technology is an innovative construction technology.

4.2.3 Case studies

Three-dimensional printing technology has been successfully applied in various sustainable construction projects, including residences, bathrooms, and high-rise buildings. The success of these projects requires the integration of 3D printing technology with creative thinking to achieve sustainable construction goals. In Asia, 3D printing technology has also been widely used. The instances listed in the following represent the application cases in Asia.

India

In 2021, India's first 3D-printed house was built in Chennai by Tvasta Manufacturing Solutions, a local start-up focused on offering low-cost housing solutions and sustainable construction (see Figure 4.2). The 55 m^2 single-storey house was constructed in just five days using squeezable concrete made of cement, sand, geopolymer, and fibre. The house includes a bedroom, a hall, and a kitchen, avoiding the complexities of conventional construction. The cost of building a 3D-printed house is approximately $10,000, which is about 30% less than traditional construction costs.

Thailand

In Saraburi, Thailand, the "world's first 3D-printed medical centre" has been constructed (see Figure 4.3). This two-storey building spans a substantial

Figure 4.3 The first 3D printed medical centre in the world made in Thailand.
Source: 3dprintingindustry

floor area of 345 m^2, making it the largest 3D-printed building among the ASEAN (the Association of Southeast Asian Nations) countries. The building material used in the 3D printer includes mortars with different strength classes for load-bearing and non-load-bearing walls. The structure's distinctive wavy walls showcase the 3D printer's capability to seamlessly incorporate "unprecedented design freedom". Moreover, the 3D-printed medical centre offers several advantages inherent to sustainable construction, including improved design flexibility, accelerated building speed, and a decreased on-site workforce requirement compared to buildings constructed using traditional construction methods. The structure was meticulously designed to withstand seismic loads, and the utilization of 3D printing technology significantly expedited the construction process.

Japan

In Japan, the "Sphere House" was 3D-printed by Serendix Partners and was first showcased in 2019 at the Fukuoka City International Tropical Horticulture Expo (see Figure 4.4). The Sphere House comprises two main parts: a spherical room with an area of about 9 m^2 and a connecting module that serves as an entrance, kitchen, and bathroom. The walls of the spherical room were primarily manufactured using bio-based plastic fibre materials through 3D printing. These materials emphasize strength, durability, insulation, heat resistance, and aesthetics, making the technology environmentally friendly.

The building features a smart control system connected to home assistant devices, allowing residents to control the house remotely via their phones or

Figure 4.4 Sphere House.
Source: 3dprintingindustry

other devices. The spherical design, with its transparent and bright window, effectively increases the sense of space. Using a concrete 3D printer, the outer walls and floors are printed in a process that can be completed within 24 hours. Supporting facilities such as water, electricity, and gas are manually installed after the main structure is finished. The entire process, from start to liveable conditions, took about three days. Furthermore, the Sphere House's spherical structure minimizes surface area, and its integrated design reduces the number of components, keeping costs low. The current cost of a 3D-printed Sphere House is about 3 million yen (around $20,000). This project explores modern architectural technology and more efficient ways to use environmentally friendly materials and renewable energy in sustainable construction.

Singapore

In 2019, the largest 3D concrete printer in Southeast Asia was installed in Singapore (see Figure 4.5). This printer can produce structures up to 9 metres long, 3.5 metres wide, and 3 metres high. For instance, it successfully printed a room measuring 3.6 m × 3 m × 2.75 m in just 13 hours. The installation of this printer cost approximately $900,000. The printer was installed at the Centre of Building Research in the Housing and Development Board of Singapore.

United Arab Emirates

The United Arab Emirates (UAE) has been instrumental in advancing the application of 3D printing construction technology. Between 2019 and 2020, the

Figure 4.5 The largest 3D concrete printer in Southeast Asia.

Source: Witteveen+Bos

Figure 4.6 The 3D printed villa in Sharjah.

Sources: 3dprinting.com

first 3D-printed villa in the UAE was constructed, taking only two weeks of printing time. The design of the building reflects a fusion of modern technology and traditional Emirati architecture (see Figure 4.6). The villa includes two bedrooms, a living room, a washroom, and a kitchen. The construction was accomplished using a robotic concrete printer mounted on caterpillar tracks.

Figure 4.7 The world's tallest 3D-printed apartment building in 2015.

Source: Winsun3d

China

In 2015, the world's tallest 3D-printed apartment building, with five floors above ground and one below, was constructed in China by Winsun Group using 3D printing technology (see Figure 4.7). The structure's components were printed into a frame structure and then assembled on-site. The building was designed for high durability, with an expected lifespan of over 50 years. Additionally, the structural components were manufactured using regenerated resources, making this construction approach a sustainable method.

Generally, compared to traditional construction technology, 3D printing technology in building construction can accurately print the required components or structures without the need for moulds. This can greatly reduce construction waste, shorten the construction period, make the construction process more sustainable, and reduce the impact on the surrounding environment. Additionally, the printing process requires minimal labour, which greatly reduces labour costs and improves construction efficiency and safety. Therefore, 3D printing will have an even greater impact on sustainable construction and make significant contributions to global environmental and social development challenges. The application of 3D printing technology makes construction more environmentally friendly, which will promote sustainable development in the global construction sector.

4.3 BIM-based sustainable construction

4.3.1 What is a BIM-based sustainable construction method?

Autodesk defines building information modelling (BIM) as "an intelligent 3D model-based process that equips architecture, engineering, and construction professionals with the insight and tools to more efficiently plan, design, construct, and manage buildings and infrastructure". BIM is a new digital tool that integrates building information throughout the planning, design, construction, management, and maintenance phases of a building's life cycle. The information is stored in a three-dimensional database that includes geometric data, professional attributes, and information about building components, as well as non-component objects such as space and motion behaviour. This digital platform facilitates the exchange and sharing of information among stakeholders in construction projects. By using BIM, teams involved in design, construction, operation, and ownership can collaborate more effectively, improving work efficiency, conserving resources, reducing costs, and achieving sustainable construction.

BIM-based sustainable construction involves integrating sustainable practices and principles into the design, construction, and operation of buildings and infrastructure through the use of BIM technology. This innovative approach emphasizes the use of digital technologies to enhance quality, safety, and cost management while minimizing environmental impact, conserving resources, improving energy efficiency, and enhancing overall sustainability throughout the project life cycle. For example, BIM-based digital mechanical and electrical management systems can optimize HVAC (heating, ventilation, and air conditioning) and lighting systems, resulting in more energy-efficient buildings that are less expensive to operate. Additionally, BIM-based monitoring systems enable real-time tracking of building performance, allowing users to identify inefficiencies and take corrective actions, thereby reducing energy consumption and minimizing waste in the construction process.

BIM comprises sufficient information interpreted directly by computer applications and supported by digital technology, facilitating sustainable construction management throughout the project life cycle. United BIM has summarized the uses and requirements of BIM in design and construction projects, highlighting 15 aspects that are mostly or partly related to sustainable construction practices (see Figure 4.8). Notably, BIM can be used in green building evaluations (e.g. LEED (Leadership in Energy and Environmental Design)) to enhance collaboration, reduce redesign efforts, track material usage and cost savings, and facilitate the creation of energy-efficient buildings. Furthermore, BIM can be integrated with other digital tools, such as 3D modelling, visualization, and simulation, enabling designers and builders to identify potential issues early and make informed decisions.

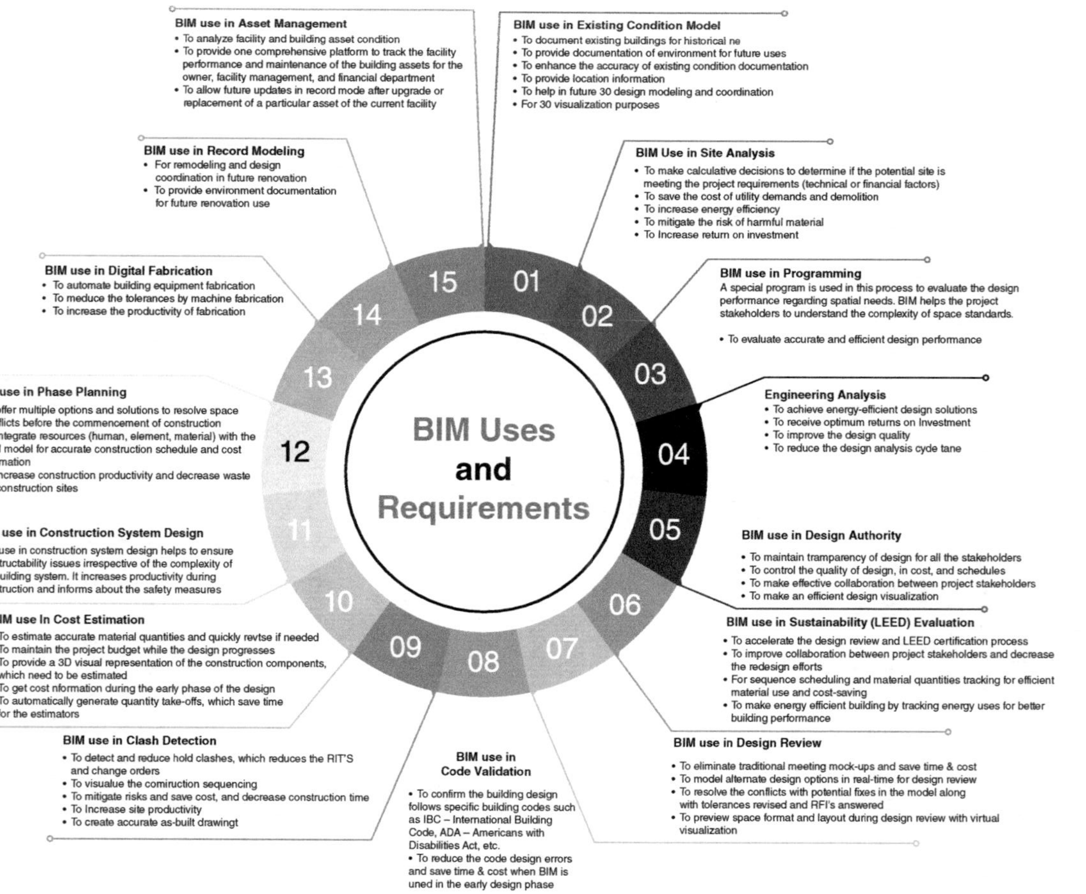

Figure 4.8 BIM uses and requirements.

Source: United BIM

4.3.2 BIM-based technology for sustainable construction

Visual and simulation

BIM technology can be used for visual checks in construction projects, allowing for the easy identification of design and construction overlaps and conflicts. Virtual and simulation analysis enables the prediction of building performance under various scenarios, guiding project teams to make informed decisions and optimize building design and operation. These simulations can also check for procedural faults due to loss or delay, and optimize net distances and layout. For example, Autodesk's "Green Building Studio" software uses BIM technology to allow users to create 3D models of sustainable buildings and conduct detailed tests and analyses on variables such as materials, energy, and environmental factors. The simulation helps architects analyse and automate infrastructure systems, including building energy models, solar arrays, LED lighting, machine algorithms, and the IoT, to achieve greater energy efficiency.

Additionally, BIM technology's virtual walkthrough function provides visual renderings of sustainable buildings, allowing users to experience the building before construction is completed. Through virtual reality technology, users can "walk inside" the building and evaluate aspects such as occupant safety, air quality, and lighting. This interactive and immersive experience fosters creativity and improvements that enhance building sustainability.

Effective collaboration

BIM facilitates collaboration among architects, engineers, contractors, and other stakeholders by providing a centralized platform to share and manage project data. This collaboration leads to better coordination, reduced errors, and improved decision-making, contributing to sustainable project outcomes. For example, the use of BIM technology for construction check can quickly and intuitively link the progress plan with the actual situation. By comparative analysis, the parties involved can effectively work together, including the designer, the supervisor, and the construction teams, and even the owners who has no engineering technical background can also have clear information and problems of the construction project.

Collision check

BIM technology can perform collision checks to detect clashes or collisions using the professional 3D information model for all parties involved in design and construction, including various design professionals, construction teams, and project stakeholders. The collision points are simulated and viewed in a 3D format, allowing technicians to intuitively understand the causes of the collisions and devise solutions. Using BIM to identify and predict potential collisions can save costs, ensure overall building safety, improve work efficiency, and help reduce unnecessary waste and environmental degradation during the building process. This early identification and resolution of potential conflicts are crucial for preventing costly rework or delays. The typical steps involved in BIM collision checks include integrating the 3D models from different disciplines, running clash detection software to identify potential collisions, reviewing the identified clashes in a 3D view, analysing the causes of the collisions, developing and implementing solutions to resolve the conflicts, and communicating the results to all relevant stakeholders.

Progress control

Progress control involves using BIM technology to monitor and manage the construction progress of a project. This process leverages the 3D models, data, and visualization and simulation capabilities of BIM software to track the status of construction activities, compare actual progress against planned milestones, and make informed decisions to keep the project on schedule. For instance, four-dimensional simulations can visualize the project plan and construction process on the key route, and important work on non-key routes can also be checked in advance. BIM can compare the actual situation with the completed work to identify existing errors. Additionally, BIM facilitates coordination between suppliers, contractors, and construction teams to ensure projects are completed on time while reducing environmental impact. By evaluating project progress through BIM, real-time construction site conditions can be monitored and adjusted to minimize waste and costs and increase efficiency.

A notable example is the Singapore Land Transport Authority's MRT station project, a large-scale infrastructure project aimed at reducing traffic congestion and improving the sustainability of the public transport system. The project was completed on time by using BIM technology to manage the design and construction processes and control progress. BIM was also used to build 3D models to simulate the environmental impacts of different design options, helping the design team optimize their choices. With the aid of BIM, the project achieved higher efficiency and quality, providing more accurate data for future maintenance, and ultimately improving sustainability. The Singapore Building and Construction Authority also published the Code of Practice for BIM e-Submission for the LTA (Land Transport Authority) requirements.

Resource saving

Using BIM for resource-saving purposes is a highly effective strategy in the construction industry. It can simulate the layout of the construction site to maximize building space and improve land use efficiency. For water conservation, BIM technology can simulate and demonstrate construction water usage for various types of equipment and stages on sites. This ensures reasonable control of water use by accounting for normal usage and losses. BIM technology also coordinates on-site water supply, drainage, and construction water to avoid waste. Furthermore, BIM can simulate energy performance and analyse the environmental impact of a building design. This allows for the integration of energy-efficient systems and designs, reducing energy consumption and carbon footprint. Additionally, BIM technology leads to material savings through scheme optimization in planning, design, and construction. It helps in collision checks, virtual construction, 3D visualization clarification, and accurate quantity estimation, facilitating rational procurement and tracking control of building materials. This approach limits excessive acquisition and reduces waste, minimizing rework and material wastage, and ultimately achieving the goal of resource conservation.

Life cycle management

BIM provides a platform for the comprehensive life cycle management of buildings. It allows for the creation, management, utilization, and revision of building information throughout the entire life cycle of a construction project. From the perspective of project management, Wang and Chen (2023) summarized BIM's capabilities across the project life cycle, from the initial design and construction phases to the handover stages, as shown in Figure 4.9. BIM enables data integration and collaboration among all stakeholders involved in the construction and life cycle management. Stakeholders can access accurate and up-to-date information about the building for various purposes throughout the project life cycle. This facilitates informed decision-making, enhances collaboration, reduces errors and rework, optimizes facility management, and increases work efficiency through streamlined processes.

4.3.3 Case studies

In Asia, BIM technology is widely used to design and manage building projects, especially in rapidly developing regions seeking sustainable development solutions. It has been employed to create 3D models and optimize designs, thereby improving the efficiency and sustainability of buildings while reducing waste, saving time, and cutting costs during the construction process. BIM technology can assess the deconstruction and recyclability of building materials, assisting designers in making informed choices to minimize environmental impact. Additionally, it can monitor a building's energy consumption and air quality, enabling architects and engineers to better manage the building by adjusting its systems accordingly. BIM technology plays a crucial role in sustainable construction in Asia, providing solutions and opportunities for the construction industry as Asian cities continue to grow.

Integrating BIM in environmental construction practices

BIM enables multidisciplinary collaboration among architects, engineers, contractors, and other stakeholders from the early design stages. This integration allows for the holistic consideration of sustainability factors such as energy

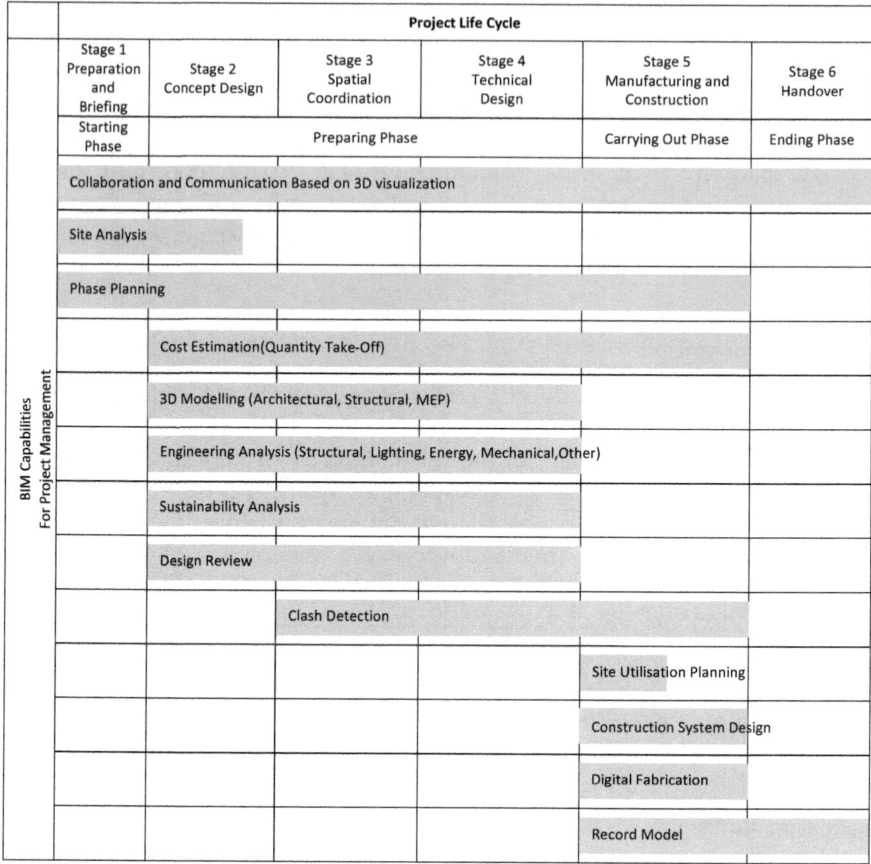

Figure 4.9 BIM capabilities for project management in the project life cycle.

Source: Wang and Chen (2023)

efficiency, material selection, water conservation, and indoor environmental quality. Additionally, incorporating BIM into sustainable construction practices allows for more informed decision-making regarding energy performance and environmental considerations, where the environmental impact of construction activities can be reduced by using renewable and recyclable materials, minimizing energy consumption, and reducing waste. Finally, BIM can be integrated with building performance evaluation and post-occupancy stages to optimize sustainability throughout the building's life cycle. By leveraging BIM technology, the construction industry can produce greener products and achieve more sustainable outcomes.

BIM technology ensures construction efficiency and minimizes environmental impact through advanced collision detection systems. For instance, during the construction of the China National Opera House in Beijing, a multi-professional 3D model collision inspection identified 89 underground and 107 above-ground collisions. This technology provides detailed information on collision locations, types, and equipment names, and proposes corresponding solutions. Implementing collision checks in BIM also improves building assembly processes, enhances design accuracy, reduces unnecessary noise and dust pollution, and improves building quality and sustainability.

BIM-based quality, safety, and cost management

BIM-based quality, safety, and cost management unifies these critical aspects in construction projects through the application of BIM technology. First, BIM automates the quality acceptance process according to relevant standards. On-site mobile terminals with positioning functions dynamically identify components, and acceptance data is automatically uploaded to a server, facilitating data sharing and collaborative work among all participants. Second, the BIM-based safety management system enables workers to wear digital safety helmets on-site, generating a virtual construction scene. Real-time location information of operators and unsafe environment data are collected via positioning functions, allowing the system to monitor and warn operators in real time. Lastly, the BIM-based cost management system links components with budget files, subcontracts, construction drawings, and schedule plans, supporting real-time calculation and statistics of engineering and subcontracting quantities based on floors, schedules, and workflows. Thus, BIM-based quality, safety, and cost management enables digital management and analysis of building information, comprehensive control, real-time collaboration, and efficient acceptance in construction project management.

In Asia, BIM-based quality, safety, and cost management has become an essential tool. It allows project participants to quickly access effective information and achieve construction goals more efficiently, enhancing the overall efficiency and quality of the construction industry. For example, the Zhuhai Dream Water City Project in Guangdong Province, China, utilized BIM technology for comprehensive quality, safety, and cost control. Dedicated BIM engineers built BIM platforms, integrating 3D models with 2D drawings, design documents, standard processes, and other information to better monitor the progress and quality of construction. During the project, all participants could share critical construction-related data and resources in a timely manner through the BIM platform, enabling close collaboration, improving efficiency, and reducing risks. The project was completed ahead of the schedule required by the owner.

BIM-based digital mechanical and electrical management system

Based on the BIM management system, a 4D model of mechanical and electrical equipment can be established to realize dynamic management and visual simulation of mechanical and electrical installation. Laser scanning, GPS, mobile communication, and other technologies are combined to track the mechanical and electrical equipment on the construction site, facilitating the management and inspection of installation progress. BIM-based digital mechanical and electrical management system is a solution that combines BIM technology with the field of mechanical and electrical engineering, providing a comprehensive solution for MEP projects. The system coordinates the relationships among various MEP equipment and systems; optimizes project design and construction processes; and enables digitized maintenance, monitoring, and management throughout the life cycle of the equipment.

An example of a successful application of this system is the Kai Tak Sports Park in Hong Kong. Covering 28 hectares, the Kai Tak Sports Park, formerly Kai Tak Airport, is the largest sports venue in Hong Kong and one of the city's most highly anticipated large-scale projects. The park features several world-class stadiums with flexible stage and seating designs to accommodate various international events and sporting activities. These venues include the main stadium, which has a capacity of approximately 50,000 spectators and is equipped with a retractable, soundproof canopy, an indoor sports arena with 10,000 seats, and a public sports ground that can accommodate up to 5,000 spectators. By employing a digital MEP management system based on BIM technology, the project achieved personalized management, data analysis, and visual display. Through modelling the system topology and process flow, the project integrated the entire chain of digital management—from design and construction to quality inspection and after-sales maintenance. This approach improved system efficiency and stability while reducing overall management costs.

BIM-based monitoring system

A BIM-based monitoring system leverages BIM technology alongside advanced technologies like big data and cloud computing to achieve real-time monitoring and assessment of building conditions. This system enhances the safety and reliability of buildings by detecting potential quality problems and structural defects promptly, allowing for timely corrective measures. By using daily construction photos, the BIM-based monitoring system generates a point cloud model of the built parts. Machine learning methods, such as vector machines, are then used to compare the point cloud model with the construction BIM model, automatically identifying progress deviations.

The Guangzhou Urban Planning Exhibition Centre in Guangzhou, China, is a successful example of this system in action. The exhibition centre is with a total floor area of 84,500 m^2. The development team implemented a structural monitoring and warning system based on BIM technology and utilized sensors, monitoring equipment, and mobile terminals to achieve comprehensive and real-time monitoring of the building's structural status. Through this BIM-based real-time monitoring and analysis system, potential safety incidents were identified and prevented before repairs, thereby enhancing the building's safety.

BIM-based green building systems

In Asia, numerous green building rating systems have been established, such as the EEWH (Ecology, Energy Saving, Waste Reduction, and Health) in Taiwan, the Comprehensive Assessment System for Built Environment Efficiency in Japan, the Green Mark certification scheme in Singapore, the National Evaluation Standard for Green Building in China, the Green Building Index in Malaysia, the Estidama Pearl Rating System in Abu Dhabi, and the Indian Green Building Council Rating System. BIM with these green building systems represents an efficient combination that enables a more effective approach to green building design and evaluation. This integration has been supported by academic research and theoretical frameworks, highlighting the synergistic benefits of leveraging both technologies to promote sustainable construction practices.

In the development of green buildings, the application of BIM is pivotal in facilitating comprehensive building performance analysis, thereby enhancing sustainable architectural design. BIM aids designers in identifying multiple alternative designs during the early stages of a project and facilitates the transmission of design information to energy analysis and simulation tools for rapid validation. Moreover, BIM provides a robust technological foundation for the storage and sharing of various types of information throughout the entire life cycle of a building.

For example, Hasanain and Nawari (2022) studied how to promote the adoption of green buildings in alignment with Saudi Arabia's Vision 2030 by developing a BIM-based application model. Vision 2030 aims to transform Saudi Arabia into an ideal sustainable society by reducing its dependency on oil and constructing more sustainable buildings and infrastructures. To achieve this, the "Mostadam" Rating System was introduced to evaluate green buildings for existing and prospective commercial and residential buildings within urban Saudi communities. The developed BIM-based assessment model aims to facilitate the implementation of the Mostadam Rating System. Autodesk Revit and Dynamo were used for programming and visualizing the model, focusing on one main category of Mostadam's Green Building Rating System: water conservation. The Dynamo plug-in can calculate water saved/reduced compared to the baseline water consumption and convert that into percentages of reduction. Based on the rate of reduction, the points earned towards the Mostadam Green Building Rating System can be calculated. Therefore, the BIM-based application model can be used in green building assessment to improve sustainability.

4.4 Industrialized construction

4.4.1 Development of industrialized construction

Industrialized construction is an innovative method that emphasizes the use of industrial production techniques to enhance efficiency and quality in construction projects. Inspired by manufacturing production lines, this approach develops standardized components, prefabricated elements, and modular systems to achieve automation, mechanization, and integration throughout the construction process. The primary goal is to reduce the time and cost associated with on-site construction while improving the overall quality and performance of building projects. The entire process of industrialized construction can be divided into four key phases: design, prefabrication, logistics, and site installation. These phases form the essential "chain" of industrialized construction, with each phase innovatively applying manufacturing philosophies and approaches.

Design phase

The design phase is crucial as it lays the foundation for the quality and adaptability of subsequent phases. Architects, engineers, and designers collaborate to create the blueprint for the construction project. Unlike traditional construction, where design is often disconnected from construction, industrialized construction requires a design process that fully considers component manufacturing and system integration. Accurate generation of executable part drawings is essential for efficient production in manufacturing factories.

Prefabrication phase

Once the design phase is complete, construction components are manufactured off-site in a controlled factory environment using machines and equipment. This includes walls, floors, roofs, and other structural elements that are then transported to the construction site for assembly. Standardized and scaled production drives overall optimization from materials and components to assembly systems. Automation and informatization in production are commonly utilized at this stage, leading to increased computer control and enhanced environmental sustainability.

Logistics and preparation phase

The logistics phase involves the procurement, storage, and safe transportation of prefabricated elements and other materials. Additionally, this phase involves preparing the land, setting up utilities, and constructing the building's foundation. Efficient logistics planning ensures that prefabricated components and materials are delivered on time and in good condition.

Site installation and construction phase

Once the prefabricated components and construction materials arrive on-site, they are assembled and installed according to the design plans. This phase requires skilled labour to ensure proper alignment and structural integrity. Other industrial construction techniques, such as lean construction and automated construction, may also be used during this phase. Throughout the construction process, quality control measures are implemented to ensure that the construction meets safety and quality standards, including structural integrity tests for prefabricated structures.

In summary, industrialized construction represents a significant advancement in the construction industry, offering the method that enhances efficiency, reduces costs, and improves the quality of construction projects. By integrating advanced design processes, prefabrication techniques, and efficient logistics, industrialized construction sets a new standard for modern construction practices.

4.4.2 Industrialized construction for sustainable construction

Industrialized construction and sustainable construction are closely related concepts that aim to enhance the efficiency and sustainability of building projects. Sustainable construction focuses on reducing environmental impacts and improving the overall sustainability of projects by integrating eco-friendly and energy-saving technologies. Industrialized construction supports these goals through several key advantages.

Efficiency enhancement

Traditional construction methods often involve sequential activities such as design, procurement, and construction, leading to time waste and increased costs. In contrast, industrialized construction separates and processes these activities simultaneously, significantly accelerating production speed and enhancing construction efficiency.

Environmental protection

Industrialized construction employs standardized components and large-scale production, minimizing material waste and environmental damage. This approach reduces the need for extensive cutting and drilling at construction sites, thereby reducing environmental impact.

Standardized production

Industrialized construction ensures the production of highly consistent products, as each component is manufactured on an identical production line. This standardization makes the construction process more controllable, reducing human errors and quality issues.

Recyclability and reusability

Traditional construction methods often involve materials and equipment that are difficult to recycle and reuse. Industrialized construction, however, incorporates recyclability and reusability into the design and production processes, allowing materials and products to be recycled and reused. This extends the lifespan of materials and products, reduces environmental costs, and minimizes construction waste.

In summary, by integrating industrialized construction methods into sustainable construction practices, projects can achieve greater efficiency, environmental protection, standardized production, and recyclability, thereby enhancing the economic and social benefits of the built environment.

4.4.3 Case studies

Industrialized construction has its roots in World Wars I and II, where the urgent need for rapid construction of military facilities and houses led to significant advancements in these techniques. This approach shifts the on-site production processes of buildings to factories, enhancing productivity and reducing waste. As an innovative building manufacturing process, it has seen significant application and development in Asia. In recent years, ongoing technological advancements and policy support have allowed countries like Japan, South Korea, and China to make significant strides in promoting and developing industrialized construction methods.

Japan

Japan is one of the earliest and most mature countries in the field of industrialized construction in Asia. Since the 1950s, Japan has been practicing and promoting industrialized construction based on the principles of "standardization" and "quality assurance". Over the years, industrialized construction has become the main trend in the Japanese construction industry, strongly supported by the government. Today, Japan has a comprehensive industry system and a certification mechanism providing efficient and reliable industrial solutions for building production. This method is widely applied in various types of buildings, such as houses, schools, and hospitals.

A representative case is the Sumitomo Fudosan Shinjuku Grand Tower in Tokyo (see Figure 4.10). This skyscraper, known for its triangular shape, was constructed using advanced construction engineering technology and efficient assembly methods. The building uses on-site prefabricated components and modular design, which reduced construction time and labour costs. Additionally, the building is equipped with energy-saving features, such as a natural ventilation system and LED lighting, addressing environmental concerns. The building's 51st floor serves as an observation deck, making it a popular tourist attraction.

China

Since 2016, China has been issuing policies and plans to encourage the production of high-quality residential buildings in factory settings. Industrialized

Figure 4.10 Sumitomo Fudosan Shinjuku Grand Tower.

Source: The Council on Tall Buildings and Urban Habitat

construction in China has rapidly developed, with technologies such as precast concrete structures, modular assemblies, and 3D printing being widely adopted. The "Special Action Plan" announced in 2017 aimed at promoting "green", "smart", "standardized", and "industrialized" construction technologies to support sustainable development initiatives. In 2019, the Chinese government released an "Action Plan to Promote the Transformation and Upgrading of the Construction Industry", setting data targets and incentives to encourage enterprises to adopt industrialized construction methods. The 2020 plan emphasized enhancing BIM technology application and improving the digital level of building design and product manufacturing.

Industrialized construction is widely adopted in large building projects across China, such as subway stations, shopping malls, hospitals, and schools. A notable example is the Ordos Art & City Museum, which features a unique GRG (glass reinforced gypsum) decorative structure (see Figure 4.11). The project employed a single-layer net shell structure for its skeleton and decorative shapes, with reliable connections between the roof steel structure, the internal main structure, and the steel structural skeleton to ensure spatial stability. A complete set of industrialized construction plans was developed to ensure smooth installation.

Figure 4.11 Ordos Art & City Museum.
Source: World Architecture News

Malaysia

In Southeast Asia, Malaysia has promoted industrialized construction since the release of the "Industrialized Housing Program" in 1996. Successful cases include assembled residential buildings, prefabricated steel apartment buildings, and schools. For instance, Broadway Malyan Collaborative Malaysia operates a large prefabrication plant capable of manufacturing steel, concrete, and wooden prefabricated components. The company is dedicated to developing industrialized construction technology to improve ecological benefits and construction efficiency.

In 2017, the Malaysian government released the Shared Prosperity Vision plan to promote economic growth and increase construction industry efficiency. The "Construction Industry Transformation Program" emphasized industrialized construction as a core infrastructure component. Many developers are now focusing on standardized industrialized construction solutions and investing in large-scale residential projects using this technology. By using prefabricated components and rapid on-site assembly, these projects meet market demand more effectively and achieve higher profitability.

4.5 Prefabricated construction

4.5.1 The scope of off-site, industrialized, and prefabricated construction systems

Traditional construction methods involve in-situ construction, where all activities and tasks occur on construction sites. This approach has been widely used globally due to its flexibility in adapting to site conditions and project control. However, it has several drawbacks, including longer timelines, higher material waste, worse environmental impacts, lower construction efficiency,

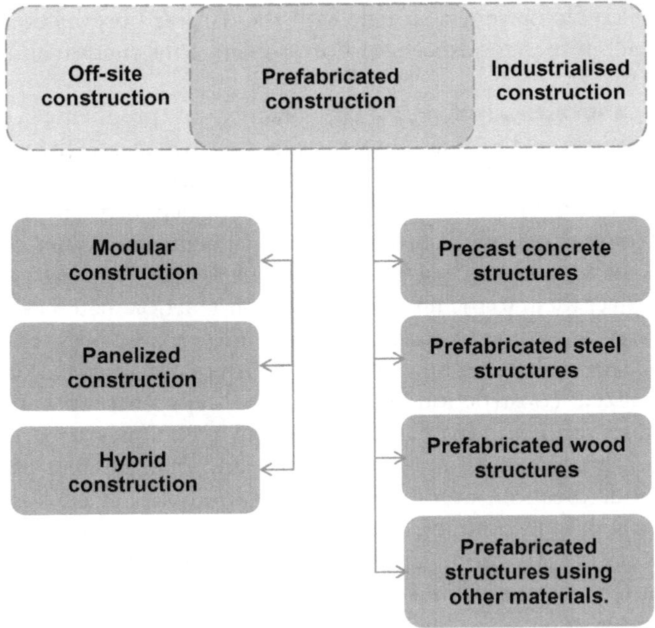

Figure 4.12 The relationship and differences among off-site, industrialized, and prefabricated construction.

and increased safety risks. In order to solve these drawbacks and meet the significant requirements of sustainable development, off-site construction has become a key method in sustainable construction. Also known as remote construction, off-site construction emphasizes the production and partial assembly of building components off-site before transporting these components to the final construction site for installation. This method generally allows components, modules, or subassemblies of construction projects to be manufactured at a location other than the construction site. These components are then transported to the site where they are assembled and constructed into the final project. Off-site construction offers significant benefits in terms of environmental impact, project control, safety, and sustainability compared to in-situ construction methodologies.

In academic research and industrial practices, terms like *off-site construction*, *industrialized construction*, *prefabricated construction*, *modular construction*, and *panelized construction systems* are commonly used to describe modern construction technologies related to prefabrication. While these terms are sometimes used interchangeably, they have evident differences, as illustrated in Figure 4.12.

Prefabricated construction is the most common advanced construction method, primarily aimed at manufacturing structural elements before their installation on projects. It involves producing building components off-site in a controlled factory environment with standardized production processes. This approach integrates aspects of both off-site construction and industrialized construction.

Off-site construction refers to the practice of producing construction elements away from the final construction site, which are then transported and assembled on-site. This method emphasizes that the production location is different from the final site, though it may not necessarily be a factory. For example, constructors might produce some elements at another construction site due to site limitations or lack of equipment, labour, materials, or methods, and then deliver them to the final site. Some construction materials, like pre-cut steels prepared according to design drawings, are prepared off-site before being transported to the final site, but this is not a form of prefabricated construction.

Industrialized construction is a broad concept encompassing all practices that use various industrial methods and techniques to streamline the construction process and improve efficiency and management performance, such as standardization, automation, prefabrication, lean construction, and just-in-time supply techniques. Prefabricated construction is one of the common industrial construction techniques, where construction elements are produced in a factory using industrial technologies, machines, equipment, and management systems.

Prefabricated construction systems can be categorized based on whether the prefabricated structures are 2D or 3D. Panelized construction involves 2D individual elements, which are then assembled into a complete structure at the construction site. Modular construction involves 3D complete elements. Both panelized and modular constructions can be integrated into a hybrid construction method. Additionally, prefabricated construction systems can be divided into different types based on the construction materials used, such as precast concrete, steel framing, timber framing, and other systems. These classifications are depicted in Figure 4.12.

4.5.2 Benefits to sustainable construction

Compared to traditional construction methods, prefabrication is a more efficient and streamlined approach where components are manufactured off-site in a controlled factory environment, then transported to construction sites for assembly. This method covers a wide range of building components and systems, such as walls, floors, roofs, stairs, and even mechanical, electrical, and plumbing systems. Prefabricated components are designed and manufactured to meet design standards, construction codes, and performance requirements such as insulation, structural integrity, and fire resistance. The benefits of prefabrication include faster construction times, improved quality control, increased safety, enhanced material efficiency, and reduced construction waste and environmental impact. Due to these advantages, prefabrication is rapidly becoming the preferred method in the construction industry globally and is likely to continue growing in popularity.

The main benefits of prefabrication technology for sustainable construction include the following.

Waste reduction

Prefabrication maximizes material usage, significantly reducing waste compared to traditional methods. Standardized components are cut and assembled in controlled environments, ensuring precise measurements and minimizing excess materials. Prefabricated construction typically has lower material waste due to these controlled conditions. Additionally, many prefabricated components are designed for reusability and recyclability, allowing them to be repurposed for new projects, which reduces the demand for new raw materials.

Energy savings

Prefabrication is known for its energy-efficient characteristics. Components are developed in factories, allowing for minimal quantities and reducing the need for heavy freight loads. Prefabricated elements are typically purchased and delivered from a single company, unlike traditional construction, which often sources materials from various locations. Using prefabricated components with energy-saving and eco-friendly features enhances energy efficiency and reduces operational energy consumption in sustainable construction projects. The controlled production environment ensures better quality control for sustainability performance.

Reduced environmental impact

Prefabrication optimizes the use of resources, including labour and materials, minimizing resource consumption and the environmental footprint of construction sites. By manufacturing components off-site, prefabrication preserves natural landscapes and reduces soil erosion and water contamination. Additionally, prefabrication significantly reduces construction time, which in turn reduces the environmental impact of on-site activities, such as noise pollution and emissions from machinery and transportation.

Improved social benefits

The factory setting for prefabrication provides a safer and more controlled environment for workers compared to traditional on-site construction. This enhances construction safety and quality, reduces exposure to hazardous conditions, and ensures the well-being of workers. Prefabrication also ensures that components meet high standards for durability and performance, resulting in buildings with longer lifespans and reduced maintenance needs. This contributes to the social sustainability of construction by extending the life cycle of buildings.

Figure 4.13 Building construction modules.
Source: Wikimedia Commons

4.5.3 Modular construction

Modular construction, also known as volumetric construction, involves the use of 3D modules prefabricated in a controlled environment and assembled on-site. This approach integrates prefabrication by manufacturing complete building units (modules) in a factory, such as bathrooms, kitchens, or guest rooms (see Figure 4.13). These modules are then transported to their final location and assembled into a complete building. The process often requires large-scale machinery and specialized workers, and incurs higher transportation costs. The advantages of modular construction include rapid construction speed, high-cost predictability, consistent construction quality, and reduced environmental impact at the construction site. In summary, modular construction is a prefabrication method that provides 3D modules, promoting sustainable construction with significant potential.

Recently, modular construction has seen significant development due to its contributions to sustainable construction. It is particularly suitable for residential and hotel projects where design standardization and strict construction timelines are essential. Companies can easily move surplus modules from one project to another, as the same modules can be produced and installed for different projects. In a factory setting, workers can produce modules more accurately with less wasted material, reducing production waste and facilitating the reuse and recycling of materials and other resources. According to a 2019 McKinsey report, modular construction could speed up construction by as much as 50% and claim $130 billion in the US and European markets by

2030, delivering annual cost savings of $22 billion. Therefore, modular construction is a sustainable construction technique that creates buildings more quickly, affordably, and with maximum resource efficiency.

Modular construction has gained significant traction in Asia in recent years due to its numerous advantages. Singapore, Japan, and China have been at the forefront of adopting modular construction practices for various projects. In Singapore, the government actively promotes prefabricated prefinished volumetric construction to address challenges such as limited space and a shortage of skilled labour, allowing for faster and more efficient construction of residential projects. In Japan, modular construction is promoted to streamline the building process, improve construction quality, and offer a more resilient solution to Japan's unique environmental challenges, particularly earthquake-resistant structures. In China, modular construction is driven by the need for rapid urbanization and sustainable development to enhance construction productivity and environmental sustainability performance. Overall, modular construction in Asia is seen as a practical and innovative approach to meeting the region's growing housing and infrastructure needs while promoting sustainability and efficiency in the construction industry.

4.5.4 Panelized construction

Panelized construction is a significant prefabrication method where building components, particularly façade systems such as walls, floors, beams, columns, and roofs, are manufactured off-site in panels (as shown in Figure 4.14). These panels are then transported to the construction site for assembly. During the process, panelized components are precisely processed in the factory, and on-site installation involves securing the panels together using bolts or other connecting methods. Panelized construction offers the advantages of prefabrication, such as improved production efficiency, enhanced safety management, reduced material consumption, and minimized environmental impacts. Compared to modular construction, panelized systems present distinct benefits in construction management by posing fewer transportation and access challenges. Overall, panelized construction is a modern and efficient method that provides numerous advantages, making it a popular choice for a wide range of building projects.

Lopes et al. (2018) reviewed the SWOT (strengths, weaknesses, opportunities, and threats) analysis of prefabricated enclosure wall panel systems compared to traditional construction methods, as illustrated in Table 4.1. Despite some weaknesses and threats to promoting the application of panelized construction techniques, the strengths present significant opportunities for future development.

Panelized construction is a widely used construction system offering the advantage of rapid assembly over traditional methods and greater flexibility than modular construction. It has been extensively adopted for residential and commercial building projects in Asia, especially as many regions transition

Figure 4.14 Panelized construction.

Sources: Atamate

Table 4.1 The SWOT analysis of the prefabricated enclosure wall panel systems (Lopes et al., 2018).

Strengths	Weaknesses
• Speed of installation (highly replicable process) reduced labour needs and costs (increase of productivity in the construction site); • Improved quality control and coordination during all stages (achieved through repetition, inspection and operating in a factory-controlled environment)-better quality of the final products; • Positive impact on supply chain management (speed of supply, reduction of work delays)-reduced costs through economies of scale; • Lower impact on the construction site (reduced time noise, waste, on-site storage area, energy costs and need for other equipment/ resources); • Increase of safety in the construction site: • Minimal maintenance • Easily replaceable/ dismantled (potential to de-construct for re-location, re-use or re-sale).	• Restricted flexibility (efficient production requires standardisation of design with relatively few variations); • Great difficulty in making changes in the construction site (requires stronger coordination since the early stages of the project); • Less used system (need for specialised workmanship); • Units may be damaged during transport and difficulty in handling larger and heavier panels (eventual need for special lifting' moving equipment).
Opportunities	**Threats**
• Adequate acoustic and thermal insulation (user comfort); • Resistance to weathering agents; • Aesthetic and functional adequacy (multiple finishes and materials available); • Flexibility of solutions, despite of the standardization of the components. • Sustainable and energy efficient buildings (reasonable energetic payback)	• Higher costs compared with traditional solutions if economies of scale are not attained (investment costs in machinery for factory production); • Lack of specific and uniformized legislation, knowledge training on the prefabricated sector. • Biased perception that prefabricated buildings/elements are of poorer quality or of smaller durability; • Choice criteria based essentially on the price; • Inertia to change by designers, contractors and users

from developing to developed economies. According to the Asia-Pacific Prefabricated Construction Market Forecast 2023–2032 (Inkwood Research, 2022), the prefabricated construction market is expected to grow at 8.96% during 2023–2032, achieving a revenue of $107.66 billion by 2032 due to the increasing number of households and the demand for new commercial and residential buildings in Asia-Pacific. In China, the prefabrication building market has significantly developed, with a 15% share in new building areas reached by 2020. In India, the prefabricated sandwich panel system, typically with expanded polystyrene, has been consistently promoted for housing projects. In Indonesia, the prefabricated market size could be US$8.47B in 2024 and is projected to reach US$11.38B in 2029, with a 6.10% growth rate during 2024–2029 (Mordor Intelligence, 2024).

Beyond modular and panelized construction systems, there is a hybrid or semi-volumetric construction system. This system combines volumetric and panelized prefabrication methods, using pre-assembled components and panels alongside traditional construction methods. It employs various components and prefabricated panels to create elements that are then assembled on-site. For example, in one construction project, bathrooms may be fully prefabricated using volumetric prefabrication, while exterior walls use panelized prefabrication. This approach allows for more flexible choices but may reduce design uniformity and is widely used in large-scale construction.

4.5.5 *Prefabricated component materials*

The materials used for prefabricated components can include various building materials such as concrete, steel, wood, aluminium, fibres, plastic, and composite materials. Among these, concrete, steel, and wood framing systems are commonly employed in prefabricated components.

Prefabricated precast concrete structure

A prefabricated precast concrete structure is a construction method where structural elements like walls (as shown in Figure 4.15), columns, beams, and slabs are manufactured off-site in a factory or production facility. These components are then transported to the construction site for assembly into the final structure. This method offers several advantages over traditional on-site concrete pouring, such as higher quality control and faster production times. Additionally, prefabricated precast concrete structures can be more cost-effective due to reduced labour and material costs. Other benefits include improved durability, increased fire and seismic resistance, and better insulation properties. These structures can also be designed to be aesthetically pleasing, with various finishes and surface treatments available. Consequently, precast concrete structures provide a reliable and efficient way to construct buildings, making them an attractive option for a range of projects, from residential homes to large commercial developments.

Figure 4.15 Prefabricated precast concrete structure.
Source: Rawpixel

Prefabricated steel structure

Prefabricated steel structures refer to buildings or structures that are constructed using pre-engineered or pre-manufactured steel components. These components include steel columns, beams, trusses, and panels that have been prefabricated off-site and then transported to the construction site for assembly (as shown in Figure 4.16). There are several benefits of constructing a building using prefabricated steel structures. First, the pre-engineered components are fabricated in a factory-controlled setting, ensuring high quality and consistency in production. Second, prefabrication reduces on-site work and can substantially reduce construction time. It also minimizes waste materials on-site, saving resources. Moreover, steel is known for its high tensile strength, making it an ideal material for buildings in areas subject to extreme weather conditions. The flexibility of steel allows for versatile designs with options for customization. Prefabricated steel structures offer excellent strength and durability and require less maintenance than traditional concrete buildings, reducing their operating costs over their lifespan. Therefore, compared to traditional steel construction techniques, prefabricated steel structures are becoming increasingly popular due to their durability, speed of construction, and cost-effectiveness.

Prefabricated wood structure

Prefabricated wood structures are buildings that are constructed using pre-manufactured wood components (see Figure 4.17). These components

Figure 4.16 Prefabricated steel structures.
Source: Flickr

Figure 4.17 Pre-manufactured wood components

include engineered wood products such as cross-laminated timber, glued laminated timber, and laminated veneer lumber that have been prefabricated off-site and transported for assembly at the construction site.

Prefabricated wood structures offer numerous advantages concerning sustainability, cost-effectiveness, speed of construction, and comfort. First, compared to traditional on-site building methods, prefabricated wood structures can reduce construction time substantially. Since the components are pre-cut and fabricated in a controlled environment, the structure can be quickly and accurately assembled on-site. Second, prefabricated wood structures are more sustainable than their concrete or steel counterparts. Wood is renewable and has a lower carbon footprint during production and transportation than other building materials. Additionally, prefabricated wood structures can offer optimal thermal insulation properties, providing excellent energy efficiency and helping to reduce heating and cooling costs. This feature may lead to the reduction of energy consumption and greenhouse gas emissions over their lifespan. Finally, this material seems to enable comfortable, healthy indoor environments due to its natural moisture regulating qualities. In summary, prefabricated wood structures will continue to gain attention and popularity with environmentally conscious builders.

CASE STUDY: THE LARGEST MASS TIMBER BUILDING IN ASIA

The largest wooden building in Asia (as shown in Figure 4.18), named the "Gaia" building, is at Nanyang Technological University in Singapore. It is a zero-energy building, comprising two six-storey curving rectangles joined at

Figure 4.18 The "Gaia" building for the Nanyang Technological University.
Source: Wood Central

multiple points. The building structure utilizes a prefabricated timber frame system, allowing for the prefabrication components manufactured off-site and reducing on-site construction time and resources. This prefabrication method reduces dust, debris, and noise pollution on sites; it is also faster and requires less labour than traditional building methods. The structure materials mainly consist of sustainable cross-laminated timber and glued laminated timber, or glulam, stiffened by a concrete core. This prefabrication technology with mass engineered timber not only benefits construction productivity and sustainability but also creates an environment conducive to learning and discourse due to timber characteristics of warmth and natural quality interiors.

There are some sustainable construction techniques applied in this project. First, the architecture team envisioned a design that fully exposed the mechanical, electrical, and plumbing (MEP) systems, intentionally omitting false ceilings to showcase these building services. Second, all design teams collaborated early in the project life cycle process to ensure the accurate production of prefabricated frame structures, which was crucial for the precise cutting, openings, and slots of the MEP components. Third, solar photovoltaic (PV) panels were installed on the rooftop, where the produced clean energy can power the usage of 169 three-room flats annually. Fourth, the wood used in the project is from sustainably managed forest, and the company for supplying the timber is fully certificated by the International Programme for the Endorsement of Forest Certification. Fifth, the prefabrication construction system was used, which can reduce waste and pollution and support disassembly at end of life to reuse and recycle. Finally, the strategy of circular economy is implemented to prioritize renewable resources, decrease reliance on finite materials, reduce waste and pollution, extend the use of assets and materials, lower carbon footprint, and regenerate natural systems.

4.6 Smart construction technologies towards sustainable construction

4.6.1 Extended reality technologies

Extended reality (XR) technologies, encompassing augmented reality, virtual reality (VR), and mixed reality (MR), are increasingly being applied in the construction sector. AR enhances real-world views by overlaying computer-generated information, while VR immerses users in a fully virtual environment. MR merges real and virtual worlds, combining the elements of both AR and VR. These innovative tools provide immersive and interactive experiences in various construction activities, such as visualizing building designs, detecting clashes, safety training, client presentations, and project coordination.

In sustainable construction, AR and VR offer significant benefits for more efficient, resource-saving, and environmentally friendly practices. They enhance visualization, improve collaboration, and optimize processes. By enabling immersive visualization of construction projects, these technologies facilitate sustainable construction practices. They improve design comprehension, facilitate early feedback, and reduce rework by allowing stakeholders to experience designs in a realistic 3D environment. AR and VR also enable the simulation of different design scenarios to evaluate sustainability factors like energy efficiency, daylighting, and material usage, optimizing sustainability performance throughout the building's life cycle. Furthermore, XR technologies support remote coordination and collaborative tools for sustainable construction. They optimize design schemes by overlaying digital models onto physical spaces, detecting clashes, and checking spatial layouts, which helps reduce waste and improve construction efficiency. XR technologies also support remote collaboration, reducing the need for physical travel and promoting a sustainable work environment. Additionally, they enhance worker skills, safety procedures, and awareness through immersive training experiences. XR technologies provide real-time feedback on building designs and construction schemes, facilitating early decisions and improving communication.

In summary, XR technologies play a crucial role in promoting sustainable construction by improving design accuracy, reducing waste, enhancing safety practices, and supporting efficient building operations. Leveraging these technologies allows construction projects to achieve environmental, economic, and social benefits, driving the transition towards more sustainable construction practices.

4.6.2 Machine Learning (ML)

ML, a subset of artificial intelligence (AI), enables computer systems to autonomously improve through experience without explicit programming. ML involves analysing extensive data sets using algorithms to learn and make predictions or decisions. In the construction sector, ML is utilized to forecast building energy usage, optimize architectural designs, enhance construction efficiency, and manage building facilities effectively.

Companies leverage ML technologies to refine construction project schedules by examining historical data to anticipate project delays and adjust resource allocation accordingly. Additionally, ML aids in designing energy-efficient management systems by analysing building energy consumption data, reducing waste, and cutting operational costs.

Another application of ML in construction involves enhancing safety protocols through image recognition and processing. ML algorithms can analyse real-time video feeds from construction sites to identify safety risks, such as workers not wearing safety gear or hazardous areas. This supports real-time safety monitoring and post-event analysis to enhance safety planning for future projects. Furthermore, by scrutinizing numerous photos and drawings, ML algorithms can reconstruct 3D models of buildings and evaluate their structural integrity, offering a cost-effective method for conserving and restoring historical structures.

4.6.3 Digital twins

Digital twins are virtual representations designed to replicate physical objects, processes, or systems accurately, primarily used for simulating, analysing, and optimizing real-time performance. In construction, digital twins simulate a building's life cycle, from design and construction to operation and maintenance. In sustainable construction, these models sustain decision-makers in understanding construction performance, predicting future requirements, and achieving more efficient sustainable construction management. Construction firms utilize digital twins to monitor construction processes, ensuring adherence to design specifications and sustainability criteria. These technologies provide real-time data on sustainability metrics like energy consumption, water usage, and carbon emissions, enabling stakeholders to make informed decisions to enhance sustainability performance, such as optimizing machinery usage and maintenance.

An innovative application of digital twin technology in construction involves managing a building's energy performance throughout its life cycle. Starting from the design phase, digital twins are employed to test various energy system designs, oversee construction progress, and continuously optimize energy utilization during the building's operational phase. Stakeholders use digital twins to gain a shared, current perspective of the project, simulating scenarios like the impact of renewable energy sources or energy-efficient technologies to determine the most sustainable options. This not only boosts the building's energy efficiency but also identifies and resolves potential issues before actual construction begins.

4.6.4 Big data

Big data involves the processing and analysis of extensive datasets that surpass the capabilities of conventional database software due to their volume, velocity, or format. In the construction sector, big data is harnessed to scrutinize market trends, refine project management, boost building efficiency and quality, and optimize energy consumption. For sustainable construction, big data plays a pivotal role in advancing eco-friendly practices by offering valuable insights, streamlining resource allocation, and enhancing decision-making throughout the construction process.

By examining vast datasets from sensors, logs, and social media, project stakeholders such as architects, engineers, consultants, contractors, and managers can gain valuable insights into sustainable construction trends. Big data analytics can predict energy usage, waste generation, and resource utilization patterns in construction projects, aiding in proactive planning and resource optimization to minimize environmental impact. Additionally, by analysing historical and real-time data, potential sustainability risks such as pollution, water scarcity, air quality, and climate impact can be proactively addressed.

In construction, big data can propel sustainable practices from a life cycle perspective. By leveraging big data in building projects, the potential for sustainable construction is maximized, enhancing both sustainability and occupant comfort. Designers can pinpoint opportunities to enhance energy efficiency and trim operational expenses by optimizing HVAC, lighting, and insulation systems. During construction, big data analytics can refine resource allocation by monitoring material usage, WHS performance, equipment efficiency, and labour productivity, fostering waste reduction, heightened project safety, and cost-efficiency while championing sustainable construction practices. Post-construction, these analytics can continue to refine project performance by identifying areas for enhancement, monitoring energy consumption, and ensuring sustained project sustainability over its life cycle.

4.6.5 Blockchain

Blockchain is a decentralized database technology that securely and transparently records information. In the construction sector, blockchain plays a crucial role in ensuring transparency, traceability, efficiency, and security in contract execution, supply chain management, and transaction reliability, all of which contribute to sustainable construction practices.

The adoption of blockchain in construction is becoming increasingly significant, particularly for enhancing supply chain transparency, reinforcing contract enforcement, and improving asset management efficiency. The inherent features of immutability and decentralization in blockchain make it ideal for ensuring transaction security, enhancing project transparency, and mitigating fraud risks. Some construction projects leverage blockchain to manage contracts, guaranteeing immutable and verifiable transaction records to reduce disputes and foster trust among stakeholders. Additionally, blockchain can track the origins of building materials, ensuring their quality and sustainability.

By utilizing blockchain technology, the carbon footprint of construction projects can be monitored by documenting data on energy consumption, material sourcing, and transportation. This information can be used to offset carbon emissions, enhance energy efficiency, and reduce environmental impact. For instance, blockchain enables a transparent and immutable record of all transactions within the construction supply chain, verifying the sourcing of eco-friendly materials, promoting recycling practices, minimizing landfill waste, tracking environmental impact, and ensuring compliance with sustainability standards.

4.6.6 Internet of Things (IoT)

IoT technology establishes connections among various devices and systems in the building environment through the internet, facilitating real-time data collection and exchange. These devices encompass sensors, surveillance cameras, lighting systems, HVAC systems, and other smart devices. In sustainable construction, IoT technology plays a pivotal role in enhancing efficiency, energy management, safety, resource utilization, and overall sustainability.

IoT technology offers real-time monitoring and control capabilities for sustainable construction practices. By monitoring energy consumption, it can optimize energy-consuming equipment, machinery, and devices to reduce energy usage and carbon emissions. Analysing performance data on energy consumption enables the establishment of predictive maintenance schedules to prevent breakdowns, minimize downtime, and prolong the lifespan of construction machinery, promoting sustainability through resource efficiency. Moreover, IoT can oversee waste generation and disposal processes, ensuring proper waste sorting and recycling, identifying areas of material wastage, and implementing measures to reduce waste. Additionally, IoT can monitor water usage, detect leaks, and optimize irrigation systems to decrease water wastage and support sustainable water management.

IoT technology enables real-time data analytics for informed decision-making in construction projects, benefiting social sustainability aspects such as WHS (workplace health and safety) performance. First, IoT devices enhance safety on construction sites by monitoring worker activities, identifying hazards, and issuing real-time alerts during emergencies. For instance, surveillance cameras and sensors monitor project security in real time, automatically detecting unauthorized access, fires, or emergencies, and promptly notifying management and emergency services. Second, by enhancing safety protocols, IoT technology safeguards workers, reduces accidents, and fosters a sustainable work environment. Lastly, by collecting and analysing data on various construction aspects like energy consumption, resource allocation, and environmental impact, stakeholders can optimize processes, cut costs, and enhance construction sustainability performance.

4.6.7 Cloud construction technologies

Cloud construction technologies utilize cloud-based platforms and services to streamline data storage, sharing, and collaboration within construction projects. These technologies offer project teams with efficient tracking capabilities and effective coordination for tasks and document management. Importantly, cloud construction technologies play a significant role in achieving sustainable construction practices by optimizing resource utilization, reducing waste, and enhancing project efficiency through the following primary pathways.

One key benefit of cloud construction technologies is the efficient storage, management, and sharing of data. These technologies centralize project data on a cloud-based platform, enabling data preparation, collection, storage, sharing, real-time communication, and collaboration among project stakeholders. This reduces paper consumption, minimizes physical storage requirements, and enhances information accessibility. Additionally, the platform allows for remote analysis and reporting of data, enabling construction teams to monitor performance, set improvement targets, and showcase environmental stewardship. By reducing reliance on physical documents, minimizing waste, and lowering carbon footprints, cloud construction technologies promote a more sustainable construction process.

Cloud construction technologies also improve sustainable construction project management practices by utilizing cloud-based supply chain management tools for efficient material procurement and rapid delivery. These technologies track and optimize material usage to ensure efficient utilization and waste reduction. Furthermore, integrating cloud-based technologies into energy management allows for real-time monitoring and optimization of energy consumption, leading to reduced energy usage and emissions. Additionally, leveraging cloud-based analytics tools enables the assessment of construction activities' environmental impact, facilitating data-driven decisions to mitigate negative effects. Finally, cloud construction platforms can integrate life cycle assessment tools to evaluate the environmental impact of construction materials and design choices. By leveraging cloud construction technologies, projects can enhance sustainability through improved data management, real-time updates and alerts, efficient resource utilization, waste reduction, compliance with environmental standards, and the creation of a healthier and more sustainable built environment.

4.6.8 Case study: integrated BIM technology with other digital construction technologies

BIM can be seamlessly integrated with various sustainable construction technologies, including prefabrication, 3D printing, XR technologies, IoT, ML, blockchain, and cloud-based construction platforms. For instance, incorporating cloud-based platforms and blockchain for data management can enhance construction collaboration, efficiency, productivity, and sustainability by updating BIM outcomes. Leveraging BIM project models, AR technology enables construction workers and stakeholders to visualize designs within the actual environment context, while VR technology offers immersive experiences for design reviews and stakeholder presentations, enhancing communication and decision-making processes. Furthermore, big data analysis based on BIM conducts thorough examinations using vast data generated by construction projects, identifying issues in building design, construction, and operation; enhancing project quality and efficiency; and leading to precise decision-making and improved building management.

The case study presented in this section originates from Teng and Chen (2022) and focuses on the New Zhengzhou Museum Project. This project incorporates modern smart and advanced construction technologies such as BIM technology, IoT, digital twins, and lean construction systems. The New Zhengzhou Museum Project stands as the largest single-building museum in China as of 2022, covering a land area of 148,000 m^2 and a total building area of 14,200 m^2. Additionally, the project exhibits substantial volume, design complexities, and construction challenges, featuring a strict construction timeline and elevated technical requirements. The glass fibre reinforced concrete curtain wall system of the project spans an area of up to 30,000 m^2, expanding to over 40,000 m^2 in total. The project encounters challenges related to the intricate double-curved curtain wall's substructure and cladding material processing, positioning, and installation, including bent columns, polyline fitting cross-members, and precise positioning of plane grid points and cross-members.

The project employs a forward-integrated BIM design strategy to optimize design, production, and operational processes, thereby enhancing construction efficiency and ensuring high-quality outcomes. Alongside the BIM model, on-site commissioning is performed and converted into 2D drawings with dimensioned nodes, which assist the workforce in processing and positioning on-site.

The project integrates radio-frequency identification (RFID) technology to enable wireless communication between readers and tags. This facilitates the identification, tracking, and data exchange of structural components throughout their life cycle, from cutting and production to quality inspection and installation. Importantly, an intelligent prefabrication factory production management system, based on BIM and RFID models, generates production task lists, enhancing efficiency in cost control, schedule monitoring, and task tracking.

The project also utilizes IoT technology to create an "Internet + smart construction site" cloud platform. This platform integrates the interests and needs of owners, designers, contractors, and subcontractors, providing project management applications. Through digital twins technology, the platform offers real-time visual communication of work processes, dynamic statistical analysis of construction data, and intelligent scheduling and supervision. This approach improves work efficiency, supports lean construction, and facilitates intelligent management throughout the construction process. The platform covers construction schedules, personnel, safety, and environmental aspects, advancing project timelines by 5% and reducing costs by 3%.

Finally, the project incorporates a BIM + 5D safety inspection system with mobile patrol points, allowing managers to use mobile devices to monitor safety supervision of major hazard sources, thereby eliminating risks and ensuring site safety. The project also embraces green and eco-friendly construction measures, including the installation of smart water and electricity meters, intensive land use, simplified site layouts, and a focus on environmental harmony and energy conservation, ultimately enhancing sustainable construction performance.

4.7 Conclusion

Smart construction technologies are pivotal in advancing sustainable construction practices in Asia. This chapter delves into the transformative potential of innovations such as 3D printing, BIM, industrialized construction, prefabrication, and digital construction practices. It demonstrates how these technologies can revolutionize construction processes by reducing material waste and labour costs, enhancing planning and design, improving collaboration, streamlining workflows, and minimizing environmental impact. By integrating these innovations, the construction process becomes more efficient, cost-effective, and environmentally friendly.

The examples and case studies in this chapter showcase practical applications of these technologies within the construction context. They illustrate the potential of 3D printing to create intricate and customized structures with minimal waste, the benefits of BIM in facilitating complex project management and reducing design and construction errors, and the efficiency gains from industrialized and prefabricated construction methods. Additionally, the integration of BIM with other digital construction technologies such as eXtended reality, ML, digital twins, big data, blockchain, IoT, and cloud technologies highlights how these advancements can drive sustainability in the construction sector. Furthermore, the chapter emphasizes that these technologies are not only catalysts for productivity and efficiency but also crucial in fostering sustainability within the construction industry. In conclusion, this chapter underscores the transformative potential of these technologies in achieving sustainable construction goals and highlights the importance of embracing these advancements for a sustainable future in the construction industry. The chapter further demonstrates the importance of ongoing research, collaboration, and innovation to fully harness these technologies' potential and ensure a sustainable future for the construction industry.

References

Hasanain, F. A., & Nawari, N. O. (2022). BIM-based model for sustainable built environment in Saudi Arabia. *Frontiers in Built Environment*, *8*, 950484.

Inkwood Research. (2022). *Asia-Pacific prefabricated construction market forecast 2023–2032*. https://www.inkwoodresearch.com/reports/asia-pacific-prefabricated-construction-market/

Lopes, G. C., Vicente, R., Azenha, M., & Ferreira, T. M. (2018). A systematic review of Prefabricated Enclosure Wall Panel Systems: Focus on technology driven for performance requirements. *Sustainable Cities and Society*, *40*, 688–703.

Mordor Intelligence. (2024). *Indonesia prefabricated buildings industry study market*. https://www.mordorintelligence.com/industry-reports/indonesia-prefabricated-buildings-industry-study-market

Teng, F., & Chen, Y. (2022). Research on the construction system of complicated form building based on smart collaboration: A case study of New Zhengzhou Museum Project. *Architectural Journal*, *2022*(3), 227–232.

Wang, T., & Chen, H. M. (2023). Integration of building information modeling and project management in construction project life cycle. *Automation in Construction*, *150*, 104832.

5 On-site construction techniques towards sustainable development

5.1 Introduction to on-site construction management

A construction site is an area dedicated to building or infrastructure development, encompassing a range of construction processes, tasks, and operations. However, construction sites also have significant negative impacts on the environment, including material consumption, carbon emissions, waste production, and pollution. The site involves frequent transportation of materials and components for activities, such as blasting, digging, levelling, site clean-up, ramming, piling, lifting components, and mixing and pouring mortar and concrete. Additionally, hazardous building materials can pose health risks to construction personnel. Construction waste and dust often contain harmful heavy metals, which can contaminate groundwater when buried, posing risks to nearby residents. As a result, achieving sustainable development on-site presents many challenges.

On-site sustainable construction management refers to the process of planning, organizing, and overseeing construction projects with a strong emphasis on incorporating environmentally sustainable practices and principles. It involves integrating sustainable construction techniques, technologies, and strategies into all aspects of construction project management, from pre-construction and procurement to construction and post-construction phases. The goal of sustainable construction management is to minimize the environmental impact of construction projects while optimizing resource efficiency and promoting long-term sustainability. The first significant work of sustainable construction site management is to establish an implementation organization. The strategies are further developed to implement sustainable construction site management. The main contents of sustainable construction site management consist of three parts: resources management, environmental management, and social sustainability commitments. The structure of sustainable construction site management is illustrated in Figure 5.1.

5.2 On-site construction management organization

To effectively manage construction projects, it is essential to establish an organizational structure tailored to sustainable construction. This structure

DOI: 10.4324/9781003202660-5

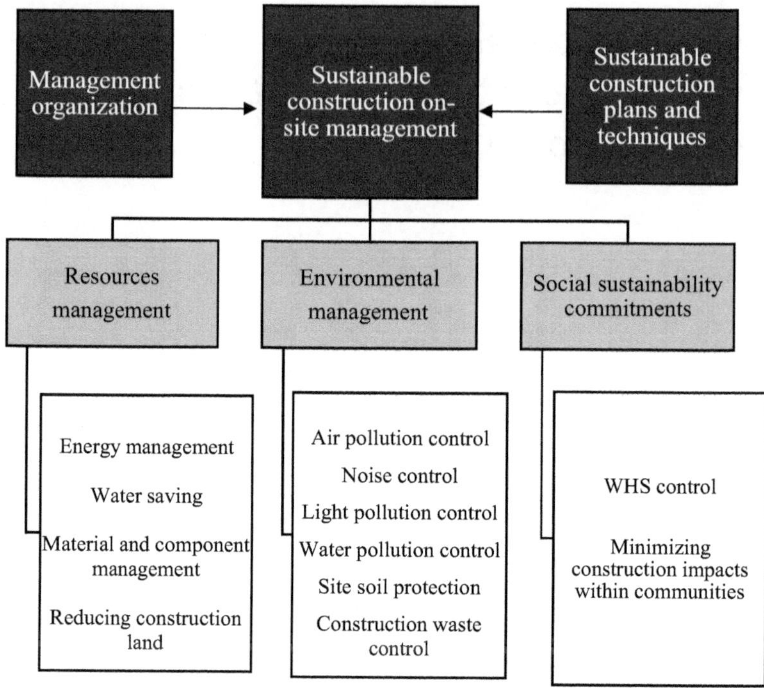

Figure 5.1 Framework of sustainable on-site construction management.

is a critical factor in the efficiency of sustainable construction management and can significantly influence project outcomes. A well-structured organizational framework helps prevent issues such as unclear accountability for sustainable construction responsibilities and barriers among team members, which can otherwise lead to project delays and disputes. Moreover, a dedicated sustainable project management team should be established to oversee the entire construction process by planning, executing, monitoring, and evaluating construction activities in a sustainable development-oriented manner. The team is responsible for developing project plans, managing project resources, and ensuring that all stakeholders are aligned and working together towards achieving sustainable construction goals. Within the team, key personnel involved in sustainable construction projects should be assigned clear roles and responsibilities to ensure effective project execution. At the very least, this includes a project manager and a site control staff. Therefore, it is crucial to have a clear understanding of each person's roles and responsibilities and how their contributions fit into the overall success of the project.

A project manager plays a pivotal role in ensuring the successful implementation of these environmentally friendly initiatives. The project manager should support the integration of sustainable development principles into project management practices. The roles and responsibilities of the project manager should consist of:

- Develop clear sustainability objectives. Work with the project team to define specific sustainability objectives and strategies, such as energy efficiency targets, waste reduction goals, and water conservation strategies. These objectives will guide decision-making and performance evaluation. Ensure that sustainability objectives are embedded into the project plan from the outset. Develop implementation strategies that outline how green practices will be integrated, tracked, and monitored throughout the project life cycle.
- Sign the contracts with sustainable development requirements. If possible, ensure that construction projects adhere to relevant sustainable development standards and certifications. Collaborate with procurement teams to source and procure eco-friendly materials and products. Verify that materials meet sustainability criteria, such as being locally sourced, having recycled content, and/or having low VOC (volatile organic compound) emissions. Coordinate the implementation of energy-efficient technologies and renewable energy solutions, such as solar panels or energy-efficient HVAC (heating, ventilation, and air conditioning) systems, to reduce the project's carbon footprint.
- Educate and engage the project team. Provide training to project team members about the importance of green construction and how their roles contribute to sustainability goals. Educate the team about green construction principles, sustainable materials, energy-efficient technologies, and waste reduction strategies. Foster a sense of ownership and commitment to eco-friendly practices. Understand the environmental practices, benefits, and challenges associated with green construction.
- Collaborate with stakeholders. Engage with the company, engineers, clients, subcontractors, and suppliers to collectively understand the goals of the green construction project. Collaborate on integrating sustainable practices into the project's design, planning, and execution. Regularly track and communicate the project's progress towards achieving its sustainability objectives. Share updates with clients, stakeholders, and team members to demonstrate the project's positive environmental impact.
- Waste management and monitoring of green construction. Develop and implement a waste management plan that emphasizes recycling and waste reduction. Conduct regular inspections to verify the use of sustainable materials and proper installation techniques. Ensure that construction waste is properly sorted and disposed of according to environmentally responsible practices. Monitor construction activities to ensure that green construction practices are being followed.
- Continuous improvement. Identify opportunities for continuous improvement in green construction practices. Encourage feedback from the project team and stakeholders to refine strategies for future projects. Anticipate and address challenges that may arise while implementing green construction

practices. Adapt project plans as necessary to overcome obstacles and ensure that sustainability goals are achieved. By taking these proactive steps, a project manager can effectively lead the company's transition into green construction, contributing to the company's commitment to environmental sustainability and delivering projects that align with eco-friendly practices.

In the construction site management team, in addition to a project manager, there should be a sustainable control staff member on the construction site who plays a crucial role in implementing environmentally friendly initiatives. This position could be called a sustainable construction engineer (SCE). By actively participating in the project's green construction initiatives, an SCE can make a significant impact on reducing environmental impacts and contribute to the successful implementation of sustainable construction practices. Here's what an SCE should do to be involved with green construction:

- Attend training and education sessions. Participate in training sessions and workshops that educate field staff about the principles and practices of green construction. Learn about specific green building practices relevant to your role. Understand the importance of sustainability and how your role contributes to the overall goals. Understand how to handle and install sustainable materials, energy-efficient equipment, and technologies that support the project's environmental objectives. Stay updated on new green construction practices, materials, and technologies. Embrace opportunities for continuous learning to enhance the skills and knowledge in sustainable building.
- Implement sustainable construction goals and strategies. Prioritize green practices and ensure sustainability on construction sites. Adhere to the company's sustainable construction goals and strategies, which may include specific procedures for waste sorting and disposal, energy-efficient installation techniques, and the use of eco-friendly materials. Follow best practices when installing sustainable materials such as energy-efficient insulation, low-VOC paints, and products with recycled content.
- Efficient resource management and environmental protection. Use resources efficiently, including materials, water, and energy. Be mindful of energy consumption during construction activities. Follow best practices for energy-efficient equipment operation. Sort and dispose of construction waste according to the company's waste management plan. Separate recyclable materials from general waste to maximize waste diversion rates. Minimize waste generation and conserve resources whenever possible. Take precautions to prevent pollution and protect local ecosystems during construction activities. Conserve water by minimizing runoff, ensuring that water sources are turned off when not needed and properly using water-efficient fixtures and equipment.
- Continuous improvement. Understand and embrace the company's commitment to sustainability. Communicate with other team members to

ensure everyone is aligned with the sustainable construction goals. Address any questions or concerns related to sustainability practices. Provide feedback to project managers about the feasibility and effectiveness of implementing sustainable practices on the construction site. Align your work with the company's values and actively contribute to its green construction efforts.
- Documentation and reporting. Maintain detailed records of the project's sustainability efforts, including material certifications, energy usage data, waste diversion rates, and any other relevant information. Prepare regular reports for internal and external stakeholders. If you encounter challenges related to green construction practices, communicate these issues to project managers. Based on the guidance from the project manager, field staffs implement how to address and overcome these challenges.

5.3 Sustainable construction plans and techniques

5.3.1 On-site sustainable construction plans

Developing plans is one of the fundamental works of implementing sustainable construction site management. It refers to the systematic process of coordinating various elements of the project, including administration, surroundings, resources, risks, communication, and adherence to project objectives. Developing and implementing plans ensure the successful execution of a green construction project. The plans should outline the specific sustainable practices that will be implemented throughout the construction process. Based on different project management structures and procedures, these plans are probably developed by SCE, reviewed by other team members, and approved by the project manager and other related stakeholders. In the context of sustainable construction, the primary management plans commonly consist of the following plans:

- Environment protection plan. It is the most common plan in construction project management, as it provides the foundation for all other plans and ensures alignment with sustainable principles and project management targets. It generally outlines the construction project's sustainability goals, objectives, strategies, key performance indicators, timelines, roles, responsibilities, and management.
- Energy management plan. It focuses specifically on achieving energy efficiency targets to minimize energy consumption and reduce greenhouse gas emissions during construction processes. It generally includes measures and demands of optimizing the energy-consumption process, using energy-efficient construction equipment and machines, applying a low-carbon supply chain process, and incorporating renewable energy sources.
- Water saving plan. Develop strategies and measures to efficiently save and use water, aiming to conserve resources and reduce consumption. This plan may include measures such as selecting water-efficient fixtures, implementing

rainwater harvesting, recycling and reusing water, adopting sustainable landscaping practices, and installing water meters and monitoring systems to ensure minimal water use.

- Consuming materials and resources plan. It aims to minimize the environmental impact of the procurement, consumption, and disposal of materials and resources. It includes key strategies such as purchasing and using sustainable materials, minimizing construction consumption, and promoting material reuse and recycling.
- Construction waste management plan. It aims to reduce the environmental impact, conserve resources, and adhere to waste management regulations. It outlines systematic strategies for reducing, reusing, and recycling construction and demolition waste. The waste sorting, disposal, and tracking waste diversion procedures should be included as well.
- WHS (work health and safety) management plan. It addresses strategies for creating a healthy and safe indoor and outdoor work environment for construction workers so as to promote the well-being and productivity of workers. It includes measures for safe work, air quality management, dust and noise control, proper ventilation, suitable lighting, hazardous substance management, risk management, and others.
- Socially sustainable construction plan. It aims to prioritize the well-being of communities, workers, visitors, and other stakeholders during construction in the surrounding areas. It generally designs measures for fostering community engagement, enhancing the overall quality of the work environment, minimizing negative construction impacts on surroundings, and creating positive social impacts.
- Monitoring and reporting plan. It aims to ensure that the construction project meets its sustainability goals and provides data for continuous improvement. The plan should include the procedures for ongoing monitoring, measurement, and reporting of construction sustainability performance indicators such as energy utilization, water consumption, waste reduction/recycling/reusing/disposal rates, WHS risks, work environment and other relevant key performance indicators (KPIs).

These plans help guide the project team in achieving environmental and social sustainability goals on construction sites. The specifics of these plans may vary depending on the project's size, complexity, and the construction business' sustainable development objectives. It is important to regularly review and modify plans to address changes or unexpected challenges that may arise from national regulations, local conditions, or stakeholder input.

5.3.2 On-site sustainable construction techniques

Effective sustainable construction site management is crucial for the successful execution of sustainable construction. The management methods are implemented to ensure that the construction progresses smoothly, adheres to sustainable practices, and stays with construction targets. Accordingly, effective implementation of the management methods can help minimize the negative construction environmental impact, enhance community well-being, and contribute to long-term sustainability on-site. The following are key sustainable construction site management methods.

- Effective communication: In sustainable construction site management, effective communication is crucial for engaging stakeholders and ensuring their active involvement throughout the project. Effective communication includes various aspects, such as establishing clear sustainable construction targets and communication channels among project stakeholders, frequency of communication (e.g. meetings and reports), and creating dedicated platforms for timely feedback and transparency. It helps the team convey messages effectively, build stronger relationships, and contribute to a positive and collaborative environment in promoting sustainable construction.
- Stakeholder engagement: Engaging stakeholders is crucial for the successful implementation of sustainability practices. Sustainable construction site management should emphasize collaboration and coordination among all stakeholders on construction sites, including local communities, designers, engineers, contractors, subcontractors, suppliers, workers, and users. It will integrate sustainability principles into all stages of the whole construction project process, from planning to execution, and enable all stakeholders to work together to achieve a common goal of sustainable development. It will be valuable to involve all stakeholders at the early planning stage to collect their concerns, perspectives, and knowledge. The key thing of stakeholder engagement is to promote open communication and mutual understanding among all stakeholders. In summary, stakeholder engagement could ensure sustainability integrated into project decisions, foster innovative sustainable practices, promote social accountability, and enhance overall project success.
- Monitoring and reporting: Regular monitoring and reporting of key sustainability performance metrics are essential for tracking sustainable construction processes and identifying areas for improvement. The KPI indicators generally include resource efficiency, energy consumption, carbon footprint, waste reduction, WHS control, continuous improvement performance, and stakeholder satisfaction surveys. Monitoring and reporting will enable the project managers to adjust as necessary to ensure that sustainable development principles are effectively integrated into construction management practices. Project managers and SCEs should first establish robust monitoring systems to collect relevant data, ensuring accuracy and reliability. Through periodic reporting, stakeholders can stay informed about the project's sustainability performance and contribute to decision-making processes. These reports should provide clear and concise information, highlighting both achievements and areas requiring attention.

- Training and development: Provide ongoing training and development opportunities for team members to enhance their knowledge and skills in sustainable construction practices. The training contents include green building certifications, energy-efficient construction methods, waste reduction strategies, social sustainability commitments, best sustainable construction practices, and sustainable project management knowledge. It will be helpful to create a plan for sustainable construction and provide education and training programs. The programs aim to ensure all team members and stakeholders understand the importance and integration of sustainable development principles into construction site management practices throughout the entire construction project process.

5.4 Resource management

With the rapid development of the world economy, the shortage of resources has become an important factor constraining economic development. Especially as the pillar industry of the national economy, the construction industry has increasingly focused on resource consumption. Effective utilization of construction resources is a key strategy for achieving sustainable construction. The construction resources discussed in this book include energy, water, materials, and construction components. Effective construction resource management involves using resources effectively, minimizing the consumption of non-renewable resources, as well as maximizing resource recycling.

5.4.1 Energy management

Energy, the most important resource, is where sustainable application is powerfully embodied through effective management. Energy consumption in construction primarily stems from machinery and equipment operation and the procurement and transport of construction materials. Several types of construction machinery and equipment are commonly used in construction. One is earthmoving machinery, including bulldozers, excavators, scrapers, and so on. The other is machinery and equipment related to foundation engineering construction, such as well-point dewatering equipment, piling machinery, sinking pipe equipment, and equipment related to concrete piles. The third is the main structure construction machinery and equipment, including concrete mixer, concrete vibrating compaction machinery, lifting equipment, vertical transport equipment, and so on. Additionally, the use of mechanical equipment leads to noise and air pollution and other environmental problems. Therefore, it is possible to reduce energy consumption by using high-performance energy-saving equipment and advanced clean energy technology, for instance, using renewable energy sources like solar panels or wind turbines to power construction office trailers or equipment.

Another aspect that needs to be considered in the construction process is energy saving during materials procurement and transportation. From the perspective of sustainable construction, it is necessary to choose the least energy-consuming process during the transportation of raw materials and equipment from the factory to the construction site. In order to achieve the lowest total energy consumption during the transportation of raw materials and equipment, the construction needs to consider two factors: one is the distance of transportation, and the other is the transportation mode. Transportation modes are normally divided into road transportation, rail transportation, sea transportation, and a combination of the former ways. In general, sea transport is more energy efficient than rail transport and road transport.

Finally, energy management should also be a focus for temporary buildings and lighting systems. Whenever possible, position temporary housing in areas with abundant natural light; ensure that the layout, orientation, and spacing of office and living spaces are thoughtfully designed to maximize natural lighting and ventilation; use energy-efficient LED lighting throughout the building, incorporate daylighting strategies to maximize natural light and reduce the need for artificial lighting during the day; and implement a maintenance schedule to ensure machinery, equipment, and systems operate at peak efficiency.

5.4.2 Water saving

Saving and recycling water resources are a crucial aspect of sustainable construction and can also help reduce construction costs. Water is primarily used for mixing concrete and cement mortar, curing soil, sprinkling and wetting ceramic tiles, and controlling dust. However, excessive emphasis on water conservation can lead to other environmental issues, such as inadequate dust suppression. Thus, it's essential to balance water saving with necessary water usage.

During the construction phase, water-saving and recycling measures can include:

- Branch water supply and metering: Implement a branch water supply system for domestic and fire control purposes, with branch water meters for measurement. Track and adjust water use each quarter and at different stages, based on actual construction needs.
- Water-saving facilities: Install water-saving fixtures such as faucets and hand-pulled flush water tanks in living, office, and toilet areas.
- Drainage and recycling systems: Set up drainage ditches around the construction site, along with circulating water pools at the inlet and outlet and collection wells for rainwater. Recycled water, after sedimentation, can be used for road and vehicle washing.
- Maintenance of water equipment: Regularly repair and maintain water-saving equipment and pipes to prevent dripping or leakage.
- Floor maintenance and conservation: Assign specific workers to manage floor maintenance, use straw curtains or plastic films to reduce water evaporation, and minimize water waste.

- Environmental protection and awareness: Enhance awareness of environmental protection and water-saving practices through effective communication and education.

5.4.3 Material and component management

The effective utilization of materials and components is an important part of environmental management. Materials on the construction site can be divided into two parts: building materials and construction materials. Building (or structure) materials are generally determined during the building design stage and consumed during the construction stage, encompassing those utilized in constructing buildings and structures, such as bricks, steel, concrete, windows, and even nails. Construction materials are used in the construction stages by construction participants, that are not part of buildings and structures, but that are used to support construction and complete the buildings and structures, such as scaffolds, formworks, ladders, disposable tools, temporary site buildings, and even tapes. For both kinds of materials, the first important pathway for sustainable construction is to choose renewable and recyclable materials. Moreover, the following pathways are important measures for promoting sustainable construction:

- Optimize cutting and sizing. Plan material cutting and sizing carefully to minimize waste. Use computer-controlled cutting systems or precision saws to optimize the use of materials. Promote standardized processing and transportation of steel bars to minimize on-site processing work, which can result in material wastage.
- Use recycled and reclaimed materials. Incorporate recycled content and reclaimed materials into the construction process. Reusing materials like reclaimed wood, bricks, or metal can reduce the need for new resources. Implement advanced modulus systems like steel, aluminium, and plastic modulus, which have longer lifespans and reduce material consumption.
- Material reuse. Salvage and reuse materials from previous projects or demolition. Reuse bricks, lumber, doors, and other materials whenever possible. Properly categorize and store offcuts and waste materials in a central location for efficient reuse, sorted by type. Use various techniques like welding, tooth jointing, bonding, and inlaying to effectively repurpose remaining and offcut materials.
- Minimizing inventory and maximizing turnover. Arrange the procurement of construction materials in stages and batches based on project progress to keep inventory levels low, avoiding excess stocks. Moreover, it is a feasible way to increase the turnover of construction materials through effective management such as lean construction.
- Efficient handling. Align the handling of materials with site layout requirements to minimize or prevent the need for secondary transportation.

Implement a maintenance system for buildings, equipment, and systems to reduce material usage during maintenance.

- Minimizing waste in temporary infrastructure. Ensure that temporary roads on the construction site align with permanent ones, reducing the need for additional materials. Choose prefabricated and reusable temporary facilities to decrease waste.
- Off-site construction method. Compared with on-site fabrication, prefabricated components have many advantages in energy saving, environmental protection, and comprehensive utilization of materials. Utilizing prefabricated components and off-site construction methods can reduce material waste and save energy during the manufacturing process. In addition, prefabricated components are also easy to recycle and reuse during the demolition.

5.4.4 Reducing construction land

Due to rapid urbanization and industrialization, many nations have experienced significant expansion in both urban and rural construction areas. Given the limited availability of land resources, prioritizing sustainable land use becomes vital for fostering human development and safeguarding the environment. Construction land reduction (CLR) plays a crucial role in ensuring the economic, ecological, and social sustainability of land use. However, as a product of new policies, CLR inevitably faces numerous barriers in meeting the diverse requirements for sustainability. For example, Zhou et al. (2022) have summarized the barriers related to the sustainable development of construction land in Table 5.1.

In order to address these barriers, it is necessary to incorporate land-use sustainability and protection strategies. This will allow construction projects to harmonize with the natural environment, conserve biodiversity, and contribute to sustainable land management practices. First, adhere to local zoning regulations and land use plans to ensure that construction activities are appropriately zoned, minimizing conflicts with surrounding land uses. Avoid disturbing the public and harming public interests. Second, choose construction sites that have minimal ecological impact and consider factors like existing vegetation, wildlife habitats, and proximity to environmentally sensitive areas. Third, conduct thorough assessments to evaluate the potential ecological impacts of construction projects before they begin. This helps identify mitigation measures and sustainable design solutions. Fourth, plan the layout to make the most efficient use of available land while minimizing environmental impact and preserving natural features. A well-planned layout enhances project efficiency, safety, and productivity while minimizing disruptions to the surrounding environment. In summary, these strategies are critical for ensuring that construction projects align with environmental and ecological principles.

Table 5.1 Summary of barriers to construction land reduction

Countries	*Barriers*
Construction land reduction in china (Zhou et al., 2022)	Lack of sustainability awareness; deficient regulations and policies; inadequate incentive mechanism; limited funding; lack of available experience; the contradiction of benefit distribution; ineffective implementation of existing regulations and policies; lack of support from subjects involved; internal conflicts at all levels of governments; incomplete information disclosure.
Brownfield development in the USA (Siikamäki & Wernstedt, 2008)	Clean-up cost; liability and longer project duration; lack of environmental assessment technique; incomplete regulatory procedure
Brownfield development in Pakistan (Ahmad et al., 2020)	Cost consideration; lack of capital; inadequate environmental law awareness; lack of collaboration between local and federal governments
Urban regeneration in Korea (Yu & Kwon, 2011)	Conflict between stakeholders; relationship between public and private interests; communication and information sharing; lack of cooperativeness of stakeholders in the project

5.5 Environmental management of sustainable construction

Environmental management is a crucial component of sustainable construction, focusing on reducing the negative impacts of buildings on the natural environment. To achieve sustainable construction, companies can incorporate designs such as low-carbon buildings and green roofs, utilize eco-friendly building materials, and optimize building layouts. Effective environmental management involves comprehensive planning and control throughout the entire life cycle of a construction project to protect the ecological environment. This includes managing waste, wastewater, air emissions, solid waste, noise, and soil disturbances.

Construction is a major consumer of resources and energy, and a significant source of pollutants. To mitigate environmental issues associated with construction, two main approaches should be considered:

- Compliance with standards and regulations: Adhering to construction standards and regulations is essential for enhancing environmental protection. This involves creating a competitive market environment that supports environmental protection, enforcing environmental laws, and establishing appropriate standards to bolster sector-wide environmental efforts.
- Internal environmental efforts: Construction enterprises should enhance their environmental practices by using clean raw materials and energy, implementing waste-derived resources, and providing green products and services.

ISO 14000 is a series of international environmental management standards established by Technical Committee 207 (TC207) of the International Organization for Standardization (ISO) in 1993. This series includes standards on Environmental Management Systems (EMS), environmental management system audits (EA), environmental labels (EL), life cycle assessments (LCA), and environmental performance evaluations (EPE), among others. ISO 14001 is specifically the certification code for Environmental Management Systems. It provides guidelines for developing and implementing environmental policies and managing environmental factors, including organizational structure, planned activities, responsibilities, procedures, processes, and resources needed for environmental assessment and maintenance.

Effective environmental protection is a fundamental goal of sustainable construction. It involves maximizing resource efficiency through energy and water conservation, using renewable materials, and safeguarding the ecological environment. This includes controlling air and water pollution, minimizing noise pollution, and managing solid waste to reduce environmental impact and preserve ecological integrity.

5.5.1 Air pollution control

Air pollution on construction sites primarily results from dust generated during material crushing and screening, as well as emissions from construction machinery. These pollutants, which include gaseous and particulate matter, can easily enter human lungs and lead to occupational diseases such as pneumoconiosis. Controlling dust is essential not only for protecting worker health and safety but also for the surrounding community and the environment. To reduce construction-related air pollution, various measures can be implemented:

- Dust control measures: Employ cleaning, sprinkling, covering, and sealing to strictly manage dust pollution at construction sites. This involves using water-based methods like hoses, misting systems, or water trucks to prevent dust particles from becoming airborne. Erecting physical barriers around dusty activities helps contain particles within the work area.
- Advanced technologies: Utilize advanced tools such as industrial vacuum systems to capture dust during activities like cutting, grinding, and drilling. Employ wet methods, like water-based cutting or grinding, to prevent airborne dust particles. Incorporate dust collection systems such as baghouses or cyclones to capture and contain airborne dust.
- Use of low-dust materials: Embrace low-dust or dust-free construction materials that produce fewer particles during handling and installation. Prefabricated concrete components, manufactured in factories, significantly reduce both dust and noise pollution compared to traditional cast-in-place concrete.
- Efficient site management: Implement efficient site management practices, including scheduling dusty activities during favourable weather conditions,

coordinating tasks to minimize dust generation, conducting regular cleaning to remove accumulated dust, and creating natural barriers like vegetation or landscaping to hinder dust dispersion. Strictly enforce bans on burning substances that produce toxic gases and harmful chemical coatings.

5.5.2 Noise control

The main sources of noise pollution at construction sites include machinery noise, transportation noise, and noise generated during production processes. Construction noise can negatively impact worker health and the living conditions of nearby residents, potentially leading to noise-induced hearing loss and other occupational diseases. Therefore, controlling noise on construction sites is a crucial aspect of environmental management.

Construction noise control can be handled from sound source, transmission route, and receiver protections. The primary approach is to manage the noise source by adopting construction methods and technologies that produce lower noise levels. This includes using fully automated and low-noise machinery for tasks that generate significant noise over extended periods. Additionally, organizing construction activities to group machines and work areas with similar noise levels together can help reduce overall noise impact. If feasible, schedule equipment usage to avoid peak work times and night-time operations to minimize noise disruption. Finally, optimize work efficiency and keep machinery and equipment well-maintained to enhance performance and reduce noise levels.

The second significant step is to block the transmission route of noise. For instance, it is necessary to build a separate temporary operating room for the machines with high noise levels by designing and using sound-absorbing materials for walls, floors, windows, and doors to provide the necessary sound transmission loss. It is advisable in design to place rooms that are sensitive to noise away from potentially noisy areas whenever feasible, both within individual units and between adjacent units. For example, the Australian Building Codes Board published the handbook titled *Sound transmission and insulation in buildings*, which shows the good and bad building layout planning examples, as shown in Figure 5.2.

The third pathway is a kind of passive control, however, which has been widely used on construction sites, such as using earmuffs or earplugs. Another pathway should be relocation, such as working at another time or location.

In Australia, the Environment Protection Authority manages noise through the Environment Protection Act and regulations. They provide procedures to prevent, minimize, or control noise and vibration impacts. The Australian Standard 2436 is a guide to noise control on construction, demolition, and maintenance sites. According to the WHS Regulation 2011, the exposure standard for noise is defined as 85 decibels (A-weighted) “averaged” over an eight-hour period, and a peak level of 140 decibels (C-weighted). The A-weighted and C-weighted are the standard weighting of the audible

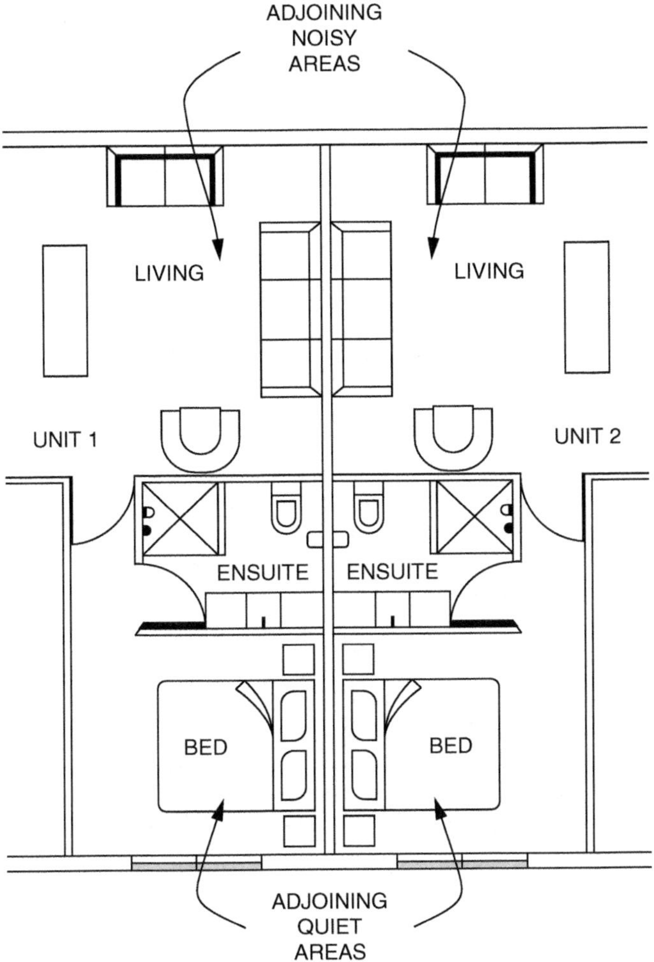

Figure 5.2 Examples of layout planning for good and bad acoustic practices (Australian Building Codes Board, 2021)

frequencies in sound level meters, which are commonly and separately used for the measurement of normal sound and peak sound pressure levels. In China, the national standard for noise pollution control at construction sites has been formulated and implemented, named Noise Limits for Construction Site (GBL2523–90). For example, pile driving is prohibited at night, and the noise of machinery and equipment is controlled under 55 decibels. Daytime noise of cranes, lifts, and other equipment should be controlled under 75 decibels, and noise of piling machinery should be controlled under 85 decibels.

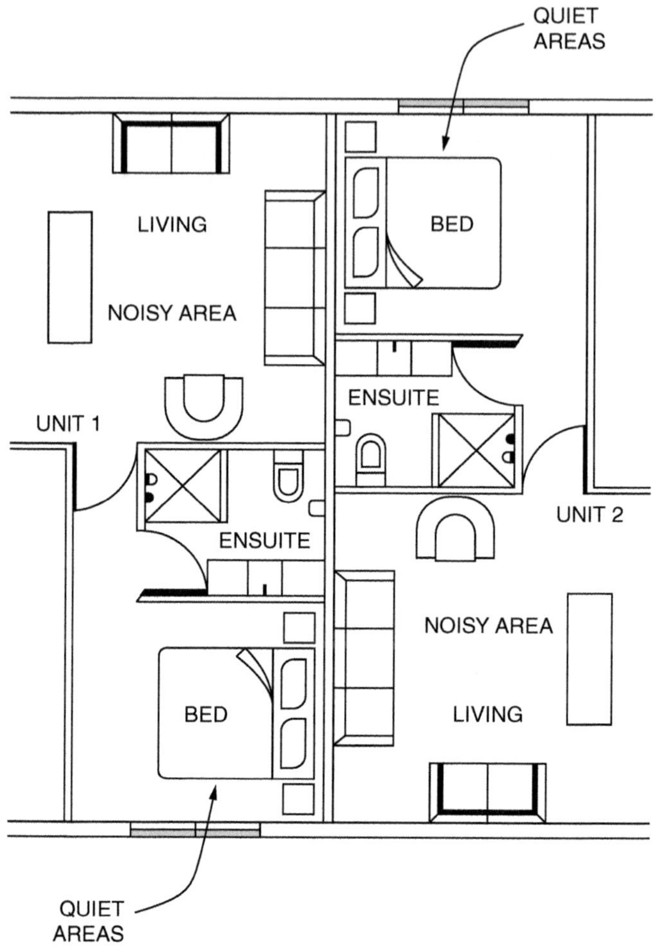

Figure 5.2 (Continued)

5.5.3 Light pollution control

Managing light pollution during construction is crucial to minimize its impact on the environment and nearby communities. Here are strategies to effectively control light pollution on construction sites:

- Opt for lighting fixtures that direct light downward to prevent spillage into the night sky. Use shielded designs to reduce glare and limit upward light emissions.
- Strategically position lighting fixtures to illuminate specific areas without excessively lighting up unnecessary spaces or extending light beyond the site boundaries.

- Incorporate timers and motion sensors to activate lighting only when necessary, reducing unnecessary illumination during inactive periods.
- Install lighting fixtures with dimming capabilities to adjust light intensity-based on-site activities and requirements.
- Focus on selective lighting, illuminating specific work areas and pedestrian routes instead of flooding the entire site with light.
- Choose warm-toned lighting with lower colour temperatures to decrease the emission of blue light, known for contributing to light pollution.
- When temporary lighting is necessary, opt for low-intensity and energy-efficient solutions to minimize light pollution.
- Implement a policy to turn off non-essential lights during periods of inactivity, such as after work hours.
- Conduct regular inspections to ensure lighting fixtures are properly directed and shielded, preventing light spillage.

5.5.4 Water pollution control

The construction industry consumes 16% of global water resources, including both direct water usage at construction sites and lesser-known indirect activities. During construction, various processes such as earthworks, dust suppression, equipment cleaning, aggregate washing, landscaping, and concrete mixing all significantly demand water resources. Besides, water pollution control in construction involves measures to prevent or minimize the contamination of water bodies—such as rivers, lakes, streams, and groundwater—due to construction activities. The main sources of water pollution in construction are wastewater and solid waste generated on-site, including mud, cement, paint, and concrete. Particularly, pollutants like sediment, chemicals, and debris can be carried by stormwater runoff into nearby water sources, causing environmental harm. Effective water pollution control strategies aim to protect water quality, aquatic ecosystems, and public health.

- Reasonable construction schemes. The schemes should be adopted to avoid or reduce the discharge of sewage and prevent water pollution as far as possible. Water pollution control plans and measures should be implemented in sustainable construction schemes.
- Erosion and sediment control. Prevent soil erosion and sediment runoff into nearby water bodies. Implement measures such as silt fences, sediment basins, erosion control blankets, and vegetative cover to prevent soil erosion and sediment runoff from construction sites into nearby water bodies.
- Proper chemical management. Handle and store construction chemicals, such as paints, solvents, and adhesives, in an environmentally responsible manner to prevent spills, leaks, and contamination of soil and water.
- Water recycling and reuse. Implement water recycling systems to capture and treat water used in construction activities, such as concrete mixing or drilling, for reuse on-site. This helps conserve water resources and reduce demand for freshwater sources.

5.5.5 *Site soil protection*

Soil is a valuable natural resource that supports plant growth, filters water, and plays a vital role in ecosystems. Construction projects can disrupt the soil structure, expose it to erosion, and introduce contaminants that can harm the environment. Effective soil protection strategies aim to preserve soil quality; minimize erosion, degradation, and contamination; and maintain the overall health of ecosystems.

- Soil testing and analysis. Conduct soil testing and analysis to understand the composition and quality of the soil at the construction site, particularly in some special situations where the land was owned by polluting or chemical companies. This information guides decisions about soil management, appropriate foundation design, and potential soil contamination.
- Preventing soil erosion during construction activities. Implement erosion control measures such as silt fences, sediment basins, erosion control blankets, and proper grading techniques to prevent soil erosion caused by construction activities. These measures help retain soil and prevent sediment runoff into water bodies. Maintain cleanliness, hygiene, and environmental friendliness, so that they will meet construction needs while ensuring a civilized, safe, and orderly site.
- Vegetation preservation and restoration. Preserve existing vegetation and trees wherever possible during construction to maintain ecosystem health and prevent habitat disruption. Particularly, implement measures to protect and conserve local flora and fauna, including sensitive habitats and endangered species, during construction activities. If vegetation is disturbed, consider restoration efforts through re-planting native species to promote ecological balance.

5.5.6 *Construction waste control*

Waste control is a critical aspect of green construction that focuses on minimizing the generation of waste and effectively managing the waste that is produced during construction and demolition processes. During the recent decades, construction waste issues have received more and more attention from both the practice communities and academics in the world due to the negative consequences on the environment and economy. Taking Bangladesh as an example, according to the latest statistics, the construction industry contributes 7.88% to the country's overall GDP and employs around 3.33 million people directly or indirectly, accounting for 5.64% of the total labour force. Globally, construction waste accounts for approximately 40% of the total waste generated, a staggering proportion. In urban areas of Bangladesh, around 16,000 tons of waste, including a significant amount of construction waste, is generated daily. Furthermore, it is projected that this number will increase to nearly 47,000 tons per day by 2025 (Datta et al., 2023). The aims of waste control are to reduce the environmental impact of waste, conserve resources, and promote responsible waste disposal. The primary management measures were planned and implemented as discussed in the following.

First, the development of effective waste management plans is crucial to facilitate the proper sorting, collection, recycling, and disposal of construction debris. The primary objective is to minimize the amount of waste sent to landfills and promote recycling initiatives and responsible disposal practices. The waste management plan should encompass detailed measures for the collection, processing, transportation, recycling, and disposal of waste materials. Additionally, waste segregation systems will be implemented on construction sites to separate various types of waste into designated containers or bins. Designated areas for waste storage and disposal should be meticulously designed, marked, and clearly labelled on the construction site.

Second, waste on construction sites should be minimized through good design strategies. By incorporating waste management considerations into the design and construction process, the amount of waste sent to landfills can be significantly reduced. Options such as designing smaller homes to minimize wastage and sourcing salvaged materials like brick, timber, and plasterboard for reuse can be explored. Moreover, sustainable construction practices can be adopted by identifying local recycling operators and opting for materials with high percentages of recycled content to contribute to the growth of recycling markets.

Third, it is essential to establish clear contracts with designers and builders and implement sustainable waste control practices. The contracts should outline waste reduction objectives and procedures for all stakeholders involved in the construction project. Waste management methods should include avoidance/reduction, reuse, recycling, recovery, and appropriate treatment or landfilling, as shown in Figure 5.3. The primary focus should be on waste avoidance and reduction wherever possible. Next, any waste that cannot be minimized

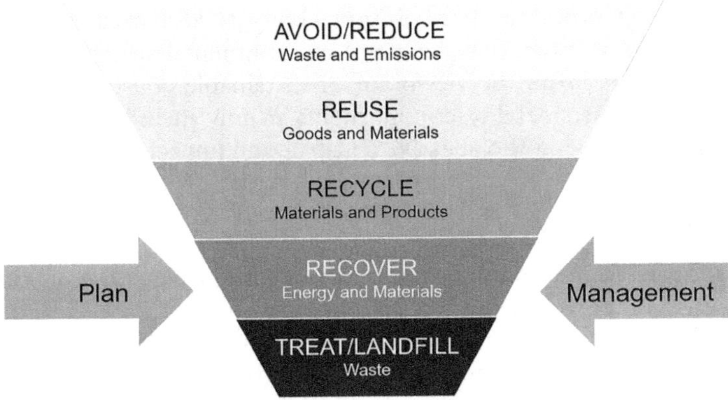

Figure 5.3 Construction waste management methods.

should be prioritized for reuse if technically feasible. Recycling should be pursued for materials that cannot be immediately repurposed. Energy recovery can be considered for waste that is not recyclable, and in compliance with local regulations, safe disposal methods for hazardous materials such as asbestos, lead-based paint, or chemical waste must be strictly adhered to.

Finally, implementing efficient construction waste management methods on-site is essential. The top priority is to establish a regular waste collection schedule to prevent waste accumulation and maintain a clean and organized construction site. Additionally, waste materials should be stored in covered containers or bins to prevent exposure to the elements. A waste tracking system should be implemented to monitor and report on the amount and type of waste generated during construction. Additionally, records of waste disposal, including the types and quantities of waste materials removed from the site and their final destination (recycling facility, landfill, etc.), should be kept. Sorting and recycling construction waste, such as concrete, metals, and plastics, can help divert materials from landfills. Collaborating with suppliers to reduce excessive packaging when delivering materials to the construction site and offering incentives to contractors and workers who participate in waste reduction efforts during construction are also recommended.

5.6 On-site social sustainability commitments

Social sustainability on construction sites involves incorporating practices and strategies of social sustainable development on construction activities, such as prioritizing the well-being of site workers and enhancing the satisfaction of the local community and surroundings. Even for the same type of construction project, social sustainability commitments may vary depending on different conditions and requirements, such as social requirements, local community, project stakeholders, and site environment. Therefore, how to commit the social sustainability in a harsh construction environment is an important part of achieving sustainable construction. Many construction companies have developed coherent Corporate Social Responsibility (CSR) strategies to amplify their positive impacts on the social and environmental sustainability systems for their business. From the viewpoint of sustainable construction site management, social sustainability commitments mainly include managing WHS on sites and minimizing the negative construction impacts on the community.

5.6.1 WHS management

Workplace well-being is an important part of sustainable construction, where prioritizing worker well-being is a foundational element for ensuring the success and longevity of projects. Worker well-being in sustainable construction is multifaceted, encompassing the safety, health, and satisfaction of workers throughout the construction process. Moreover, the sustainable safety challenge is one of the primary requirements for sustainable construction in Asia.

Therefore, the most important issue in improving workplace well-being is to manage WHS on construction sites.

WHS management is to prioritize the health and safety of construction workers by providing a safe and ergonomic working environment. WHS control is a crucial aspect of construction management which becomes even more important to ensure the physical and mental health of employees and workers in sustainable construction. It involves the implementation of high safety standards and health measures to mitigate the risks of accidents and occupational diseases. In the process of construction, effective measures should be taken to protect the health of construction workers. Methods for WHS improvement are diverse. By summarizing the best practices and experiences in the Australian construction sectors, Zhu et al. (2023) summarized the support, push, and pull measures to improve safety performance in the construction field, as shown in Figure 5.4.

Support measures

The support measures primarily focus on a "management" perspective. At the outset of a project, comprehensive well-being policies should be established to clearly define workers' rights, safety standards, and welfare benefits. Creating a collaborative environment by involving employees in decision-making can boost their pride and help in effectively identifying and addressing potential issues. To protect workers' health and reduce exposure to toxic substances, it's essential to use eco-friendly and safe construction materials and components. Effective management practices must also be implemented to maintain a clean and hygienic work environment.

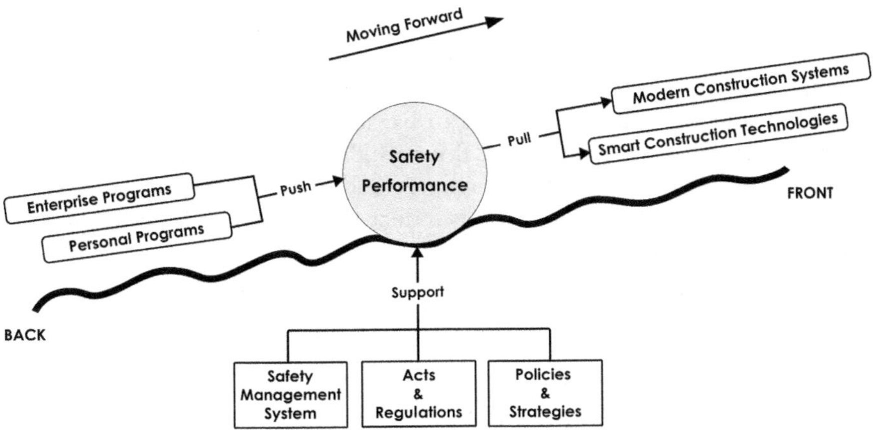

Figure 5.4 Safety performance improvement in the construction field (Zhu et al., 2023).

ISO 45001 and related standards aim to improve employee safety, reduce workplace risks, and create better, safer working conditions. Construction organizations must rigorously enforce safety responsibility systems and ensure that all workers are aware of their safety duties. Comprehensive safety education is crucial, including training for new workers, sessions before and after holidays, and refresher courses upon return to work. This education helps enhance worker safety, prevent accidents, and reduce occupational hazards. Given the dangerous and physically demanding nature of construction work, coupled with an aging workforce and global labour shortages, the industry needs to transition to intelligent, digitalized operations and move away from labour-intensive practices.

"Push" methods

The "push" methods primarily focus on education and training. Providing education, training, and development opportunities is crucial for equipping workers with the skills needed to navigate new technologies and embrace sustainable construction methods. The opportunities include proper training, protective gear, and accident prevention measures. Such investment not only boosts professional competence but also enhances worker dedication and satisfaction. Ongoing training ensures that workers stay updated on the latest safety and environmental standards, allowing them to adapt to new construction technologies and sustainable practices. By investing in their professional development, workers become more resilient to industry changes and better prepared to contribute to sustainable practices.

"Pull" methods

The "pull" methods focus on adopting modern technologies to improve both construction efficiency and worker experience. Technologies such as smart building management systems and construction robots can significantly enhance safety and efficiency. For instance, construction robots can autonomously handle physically demanding or hazardous tasks, reducing the risk of injury to workers. An example is the floor grinding robot, which eliminates the need for workers to push heavy machinery and generates less dust and noise. Similarly, exterior wall spraying robots reduce the risk of falls from heights and offer stable, controllable spraying.

One notable example is the use of VR technology, which is crucial for enhancing WHS. VR can create realistic simulations of construction environments and scenarios, providing immersive safety training. By integrating virtual reality with real-life situations, workers can experience and navigate potential hazards in a controlled virtual setting. For instance, VR can simulate scenarios such as high-fall risks or object strikes, giving workers a realistic grasp of these dangers. This type of experiential training enhances safety awareness and effectiveness, making VR a valuable tool for improving safety on construction sites.

Figure 5.5 VR training: the secret weapon to immersive construction safety education.
Source: Scratchie (2023)

5.6.2 Minimizing negative impacts on the community

When construction activities commence at a site, they can significantly affect the surrounding area, impacting the comfort, health, and well-being of nearby residents. For example, construction noise, air pollution, and disruptions to routine activities can impact the mental and physical well-being of residents, potentially leading to stress and other health-related issues. Here are some common impacts that can occur when a construction site begins its operations:

- Noise and vibrations: Construction activities involve the use of heavy machinery, equipment, and tools that generate noise and vibrations. This can lead to increased noise levels in the vicinity, disrupting the peace and quiet of the surrounding area.
- Environmental impact: Construction sites can affect local ecosystems, water bodies, and vegetation. Runoff from construction sites can carry pollutants into nearby water sources if proper erosion control measures are not in place. Construction activities can generate dust and particulate matter, which can degrade air quality and potentially lead to respiratory problems for nearby residents. Figure 5.6 shows an example of steel boundary walls commonly used in China.
- Traffic disruption: Construction vehicles and equipment entering and exiting the site can cause traffic congestion, especially if proper traffic management measures are not in place. Roads or pathways near the construction site may be closed or restricted, affecting the convenience of residents and businesses.

Figure 5.6 Steel boundary walls commonly used in China.

- Visual changes: The construction site itself can alter the visual landscape of the area, often appearing as an unsightly disruption until the project is completed.
- Safety concerns: Construction sites can present safety hazards, for both passers-by and properties. Fencing, signage, and safety protocols are essential to mitigate these risks.
- Community disruption: Construction activities might disrupt the daily lives of nearby residents, including disruptions in sleep patterns due to noise, changes in commuting routes, and overall disturbance. Construction activities might reduce opportunities for outdoor activities or social interactions, as residents may avoid areas close to the construction site.
- Property values: While property values might increase after the completion of a construction project, the immediate impact can be negative due to disruptions and changes in the local environment.
- Communication and transparency: Effective communication between the construction company and the local community is crucial to managing expectations, addressing concerns, and keeping residents informed about the progress and potential disruptions.

It's important to note that many of these impacts can be managed and mitigated through careful planning, adherence to regulations, and consideration of the needs and concerns of the local community. Minimizing the negative impacts of construction on surrounding well-being requires proactive planning, effective communication, and the implementation of various mitigation strategies. Construction companies often implement measures to minimize negative impacts and foster positive relationships with the surrounding area. Here are some approaches to consider:

- Pre-construction planning. Conduct thorough environmental assessments to identify potential impacts and develop strategies to mitigate them, such as using expert review and brainstorming methods. Designing suitable construction methods is important to minimize noise, dust, and disruption. Ensure that construction sites are visually appealing and considerate of the local surroundings. Incorporate landscaping and beautification efforts into the construction plan.
- Community engagement. Communicate with the local community early in the planning stages to inform them about the project, its timeline, and potential disruptions. Provide clear signage indicating construction zones and detours to minimize confusion. Display contact information for inquiries and feedback. Seek feedback and address concerns to build positive relationships with residents. Offer support to affected businesses through multiple reduction measures to mitigate potential revenue losses during construction.
- Traffic management. Develop a comprehensive traffic management plan to minimize congestion and disruptions. Provide alternative transportation options for workers and residents.
- Safety precautions. Implement strict safety protocols to prevent accidents and protect both workers and the community. Erect appropriate barriers and signage to ensure public safety.
- Green construction practices. Adopt sustainable building practices to reduce the environmental footprint of the construction process. Use eco-friendly materials and methods that enhance the environmental performance of the community.
- Monitoring and reporting. Regularly monitor noise, air quality, and other potential impacts to ensure that mitigation measures are effective. Maintain open lines of communication with regulatory agencies to stay compliant. Stick to the planned construction timeline to reduce the duration of disruptions and negative impacts.
- Post-construction restoration. Restore the site to its original or improved condition after construction is completed. Plant trees, shrubs, and greenery to enhance aesthetics and biodiversity.

In summary, achieving social sustainability on construction sites is an important part of promoting sustainable construction. Balancing the needs of the community, businesses, and construction projects is a complex task. It requires proactive planning, clear communication, and a commitment to minimizing disruptions while ensuring the safety and well-being of workers and residents. A well-executed sustainable construction management can help achieve construction project targets, maintain worker well-being, and satisfy the local community.

5.6.3 Case studies

Case study 1: IoT-based application in construction safety monitoring in Hong Kong

In Hong Kong, construction workers undergo training in four different types (W. Chung et al., 2023). Each of these four safety training types is expected to yield different outcomes, thus, the duration of each training varies. However, several researchers in the past have argued that the duration of training and work experience do not have a direct correlation with construction safety. Additionally, there can be overlaps in the content of the four types of training. These modular types have proven beneficial for past construction workers with lower educational backgrounds. Nowadays, with attractive salary packages, more educated individuals are entering the construction industry. These individuals often possess common expertise in the safety-related issues covered by the training. This reduces the attraction and effectiveness of traditional training, making the development of effective safety promotion increasingly important. In Hong Kong, the most effective monitoring method for checking the personal protective equipment (PPE) of each worker is the installation of booth scanners at the entrance of construction sites; workers without proper PPE are denied entry. Another monitoring method is the routine on-site inspection by safety supervisors and penalties for workers not using PPE correctly. Due to the large area of most construction sites in Hong Kong, main contractors and subcontractors do not employ sufficient safety supervisors to monitor all areas of each site in real time (Chung et al., 2023).

According to W. Chung et al.'s study, construction sites in Hong Kong apply the design of the IoT model, which aims to ensure every worker in a designated area wears PPE and trigger alerts at the on-site office in case of improper use for real-time monitoring. Furthermore, data collected from the system can be automatically stored in a database, eliminating the need for paper records from on-site safety supervisors. These data can be utilized to improve the design of safety training courses and the IoT system itself in the future. In terms of real IoT implementation, the Hong Kong Convention and Exhibition Centre (HKCEC) upgraded its building management system in 2017 to include an IoT network for collecting data on temperature, humidity, leaks, and indoor air quality. Over 500 IoT sensors were installed and connected to a wireless IoT gateway. This is a significant IoT project in Hong Kong that addressed many installation constraints, costs, and issues related to the effectiveness of IoT sensors. Building on the concept of the HKCEC IoT system, the current research examines its application in enhancing construction site safety and proposes an innovative design model for a real-time safety monitoring system using IoT technology and sensors.

Case study 2: colour settings in WHS in China

In China, safety risks are categorized according to the "Classification for Casualty Accidents of Enterprise Staff and Workers" (GB 6441–1986). This classification considers factors such as causative substances, accident triggers, hazardous materials, and harm modes. Safety risks are assessed using various methods, with the aim of preventing major and catastrophic accidents. The classification system identifies risks as major, significant, general, or low, denoted by the colours red, orange, yellow, and blue, respectively.

Following national standards, the Shenha highway project of China Railway No. 2 Bureau has developed a "Four-Colour Diagram" method to implement safety measures on the construction front line, as detailed in Table 5.2. The "Four-Colour Diagram" (officially known as the Four-Colour Safety Risk Spatial Distribution Diagram) is a systematic approach to risk management in construction projects. This approach helps prioritize safety measures, ensuring that high-risk areas receive appropriate attention and resources to mitigate potential hazards. The process involves several key steps:

Risk identification and assessment: At the project's inception, a team of experts is assembled to identify and assess safety risks. This involves reviewing project schedules, special construction plans, and third-party risk assessment reports. The team compiles a comprehensive list of potential risks and categorizes them by risk level, resulting in a "Project Construction Safety Risk Statistics Table".

Monthly risk assessment: Each month, the project team evaluates various construction site areas—such as excavations, foundations, piers, and in-situ boxes—based on their risk levels. Areas are colour-coded according to their risk severity: red for high risk, purple for significant risk, yellow for medium risk, and green for low risk.

Risk management and monitoring:

- Red and purple areas: These areas represent high safety risks and require close monitoring and management. Project leaders and the safety and environmental department must be actively involved, with designated personnel overseeing and coordinating activities in these areas.
- Yellow areas: These are medium-risk zones that necessitate regular monitoring and inspections by project leaders and the safety and environmental department. Relevant managers and teams should be present to supervise and coordinate.
- Green areas: Representing low-risk areas, these require routine maintenance and inspection.

Table 5.2 Colour settings in WHS in China (Construction Enterprise Management Magazine, 2021)

Colour	*Risk level*	*Requirements*	*Sections*
Red	Very high	During construction, the corresponding management personnel must be present, and the safety and environmental department must closely supervise, while the project leaders closely monitor the situation. Special personnel within each team should be assigned to direct and coordinate the activities.	Excavation for foundation pits with a depth exceeding 5 m, construction of pier support formwork beams with a height above ground exceeding 15 m, construction of movable formwork beams, and construction near high-voltage power lines, among others.
Purple	High	During construction, the corresponding management personnel must be present, and the safety and environmental department must closely supervise, while the project leaders closely monitor the situation. Special personnel within each team should be assigned to direct and coordinate the activities.	Excavation for foundation pits with a depth of 3–5 m, construction of piers with a height above ground ranging from 8 to 15 m, construction of bridge railings, crane lifting operations, construction near underground pipelines, among others.
Yellow	Medium high	During construction, the corresponding management personnel must be present, and the safety and environmental department must closely supervise, while the project leaders closely monitor the situation. Special personnel within each team should be assigned to direct and coordinate the activities.	Pile foundations, foundation pits for which the excavation depth is less than 3 m, piers with heights above ground less than 8 m, bridge deck systems, etc.
Green	Low	Completed areas; safety zones without electrical wiring and various types of pipelines.	

5.7 Conclusion

Chapter 5 highlights the critical nature of on-site construction techniques in fostering sustainable development within the construction sector. The chapter delves into the multifaceted aspects of on-site management that contribute to sustainability, showcasing the pivotal role that effective management practices play in achieving environmentally, socially, and economically sustainable construction projects. It underscores the significance of establishing structured and efficient management frameworks, outlining the roles and responsibilities of key stakeholders, and developing comprehensive management plans to guide construction activities. The chapter also emphasizes the importance of optimizing resource usage, implementing effective environmental management strategies, and prioritizing the well-being of construction workers and the surrounding community. By adopting these practical on-site construction techniques, construction management organizations can pave the way for promoting the sustainability of construction projects. In this chapter, the insights gained from exploring on-site construction management practices underscore the need for ongoing research and improvement to fully realize the potential of sustainable construction outcomes. The valuable insights and practical solutions that can be implemented in real-world projects will contribute to a more sustainable future for the construction industry.

References

Ahmad, N., Zhu, Y., Lin, H., Karamat, J., Waqas, M., & Mumtaz, S. M. T. (2020). Mapping the obstacles to brownfield redevelopment adoption in developing economies: Pakistani perspective. *Land Use Policy*, *91*, 104374.

Australian Building Codes Board. (2021). *Sound transmission and insulation in buildings*. www.abcb.gov.au/sites/default/files/resources/2021/Handbook_Sound_Transmission_and_Insulation_in_Buildings.pdf.

Chung, W., Tariq, S., Mohandes, S., & Zayed, T. (2023). IoT-based application for construction site safety monitoring. *The International Journal of Construction Management*, *23*(1), 58–74.

Construction Enterprise Management Magazine. (2021, March 23). *Safety production has a smart "four-color map" method to guide (In Chinese)*. https://mp.weixin.qq.com/s/gxyRH36g6374B013_ouq2Q

Datta, S. D., Rana, M. J., Assafi, M. N., Mim, N. J., & Ahmed, S. (2023). Investigation on the generation of construction wastes in Bangladesh. *The International Journal of Construction Management*, (13/14), 23.

Scratchie. (2023). *VR training: The secret weapon to immersive construction safety education*. www.scratchie.com/post/vr-training-the-secret-weapon-to-immersive-construction-safety-education.

Siikamäki, J., & Wernstedt, K. (2008). Turning brownfields into greenspaces: Examining incentives and barriers to revitalization. *Journal of Health Politics, Policy and Law*, *33*, 559–593.

Yu, J.-H., & Kwon, H.-R. (2011). Critical success factors for urban regeneration projects in Korea. *International Journal of Project Management*, *29*, 889–899.

Zhou, J. H., Zhu, Y. M., He, L., Song, H. J., Mu, B. X., & Lyu, F. (2022). Recognizing and managing construction land reduction barriers for sustainable land use in China. *Environment, Development and Sustainability*, *24*, 14074–14105.

Zhu, R., Hu, X., Wei, A., Yang, W., & Ji, F. (2023). Measuring safety performance of construction employees using data envelopment analysis: A case in Australia. *Journal of Safety Research*. https://doi.org/10.1016/j.jsr.2023.11.016.

6 Introduction to sustainable construction materials

6.1 Introduction

The construction industry is a major consumer of natural resources and significantly impacts the environment. According to the UNEP, the global construction sector accounted for over 34% of global energy demand and around 37% of energy and process-related CO_2 emissions in 2021. The World Green Building Council highlights this further, estimating that the construction industry uses 40% of the world's raw stone, gravel, and sand resources. Additionally, the construction phase is responsible for 20%–50% of total carbon emissions throughout a building's life cycle, with an average annual emissions ratio of 0.62 between construction and operation (Chen et al., 2023). These statistics emphasize the significant impact of the construction industry on natural resource consumption and environmental sustainability.

Due to the increasing pressure of climate change and the finite nature of resources, the construction industry must address these sustainability challenges. Sustainable construction aims to mitigate environmental impacts by using renewable and recyclable materials, reducing the embodied energy in building materials, lowering the energy consumption of completed buildings, minimizing on-site waste, and protecting natural habitats during and after construction. Using sustainable, renewable, recyclable, and reusable materials is essential for practising sustainable construction. Therefore, this chapter will investigate and promote green building materials, recycled and reused materials, and waste materials in construction, aiming to highlight the critical need for the construction industry to adopt more sustainable practices.

6.2 Construction materials versus building materials

The terms "construction materials" and "building materials" are often used interchangeably, but they can have different definitions and scopes depending on the context. "Building materials" refer to the materials that broadly encompass them and are used for the components of buildings or structures, such as wood, concrete, steel, glass, bricks, insulation materials, roofing materials, and more. They are substances or products used in building buildings

DOI: 10.4324/9781003202660-6

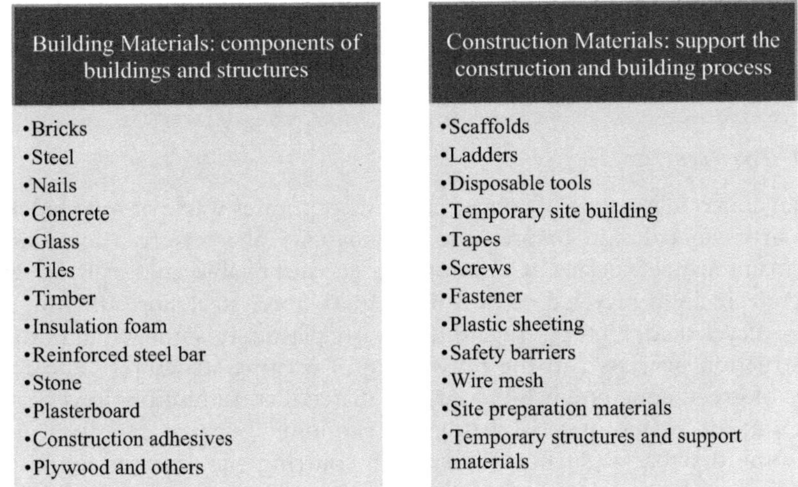

Figure 6.1 The examples of building materials and construction materials.

and other structures. For the requirements of sustainable development, building materials are generally defined as "green". Green building materials, also known as sustainable or eco-friendly building materials, are materials that have been selected and designed with environmental considerations in mind.

"Construction materials" are the materials specifically used in the actual construction process to support the construction of buildings and structures, such as scaffolds, ladders, temporary site buildings, safety protections, and others. These construction materials contribute to the overall construction process, ensuring that the site is prepared, safe, organized, and equipped for the construction activities that will take place. They support the building process by providing the necessary infrastructure and resources for workers and equipment. For the requirements of sustainable development, construction materials are generally required to be "recyclable and reusable". The differences and examples of building materials and construction materials are demonstrated in Figure 6.1.

6.3 Green building materials

Commercial and residential facilities account for approximately 40% of global energy-related carbon emissions. Constructing green buildings is an effective way to reduce these emissions. It is essential to achieve a high level of efficiency in reducing energy, water, and material consumption for green buildings. The primary objectives of green building construction methods are to reduce resource depletion and waste generation in the building industry. One primary strategy for constructing green buildings is using environmentally friendly materials that enhance the quality of the building environment. These materials are expected to be natural, durable, reusable, and recyclable. The following are some green building materials commonly used in construction practices.

6.3.1 Green concrete

Green concrete is a type of concrete that incorporates waste or recycled materials in its mix design to promote sustainability and ensure a long-lasting, low-maintenance surface. Commonly used sustainable materials in green concrete include recycled carbon fibre, glass fibre, steel fibre, fly ash, silica fume, recycled concrete aggregate, and waste plastic. It is utilized in both new construction projects and the retrofitting of existing structures. The advantages of green concrete include reduced material consumption, lower carbon footprint, decreased material waste, and economic benefits. However, it also has some drawbacks, such as difficulty in sourcing the required amounts of materials, high water absorption, and lower compressive strength for certain materials. Despite these challenges, green concrete significantly contributes to the sustainability of construction projects by reducing their carbon footprint.

6.3.2 Green thermal insulation

Insulation is a crucial factor in a home's energy efficiency. Homes with inadequate insulation require more energy for heating and cooling, leading to higher costs. Proper insulation materials, ideally eco-friendly, are designed to minimize energy consumption. Heat naturally moves from warmer areas to cooler areas, so the purpose of insulation is to resist this air movement. Consequently, insulation is necessary for roofs, ceilings, walls, and floors to prevent air from entering or escaping, as air can move in multiple directions.

Insulation effectiveness is measured by its thermal resistance value. The better the insulation, the higher the value. The value is determined by the type of insulation, its thickness, density, installation location, and other factors. Safe and suitable insulation materials should have lower manufacturing costs, such as the materials discussed in the following.

- Natural formaldehyde-free materials such as wool, cotton, aerogel, denim, thermal cork, polystyrene, cellulose, and icynene.
- Glass wool—recycled glass materials. Glass wool is predominantly made from recycled glass products, including flat glass sheets and glass bottles. The remaining glass comes from sand, which is a plentiful resource. Glass is broken down and spun into fibres that are knitted together to create rolls or batts of glass wool. If glass wool is left intact, it can be recycled at the end of its useful life.
- Polyester—recycled plastic materials. Polyester batts come from recycled plastic materials such as plastic bottles. Polyester may be recycled at the end of its useful life, just like glass wool. Because polyester insulation does

not contain any permeable particles, it is a great option for those who have allergies or asthma.
- Rockwool—crushed rock materials. Rockwool is made from crushed rock particles and stone dust, which can include up to 95% recycled resources. The exceptionally high fire ratings of rockwool insulation will aid in containing the spread of fire.

6.3.3 Structural insulated panels

Structural insulated panels (SIPs) are one of the most airtight and well-insulated building technologies available. SIPs are composed of a rigid foam core of insulation sandwiched between structural skins. Different core materials can be used, including expanded polystyrene, polyurethane, polyisocyanurate, and extruded polystyrene. All these materials provide high levels of insulation and are entirely consistent within the panel, interrupted by minimal framing lumber. SIP skins are commonly made of oriented strand board, although for particular uses, they can also be composed of other materials.

The major components of SIPs, foam, and oriented strand board, require less energy and raw resources to manufacture than conventional structural construction systems. In addition to being manufactured in a controlled environment, SIPs are more productive than site-built frameworks. Moreover, building with SIPs creates a superior building envelope with excellent thermal resistance and less air intrusion, enabling improved control over indoor climatic conditions. In summary, the most significant environmental benefit of SIPs is reduced energy use for heating and cooling throughout the life of the building.

6.3.4 Reclaimed wood

Wood is a major construction material that is sustainable and renewable. Reclaimed wood can be used in green buildings for both residential and commercial fields. For example, reclaimed hardwood floors, decks, wall panelling, tables, countertops, cabinets, shelving, and any other woodworking projects are all possible with this lumber. Reclaimed wood has attracted more and more attention in the construction field, and has the following advantages.

- Environmentally friendly: By opting for salvaged wood, the demand for freshly harvested timber is reduced, thereby mitigating the impact of deforestation. Reclaimed wood, when responsibly sourced, serves as a sustainable alternative, minimizing both landfill waste and the reliance on environmentally harmful manufacturing processes. Therefore, reclaimed wood contributes to environmental conservation, helps preserve natural resources, and promotes a more circular and sustainable approach to resource utilization.
- Durability: Reclaimed wood generally has greater strength and durability compared to its conventional solid counterpart. Having already undergone

prior use and exposure to the elements, reclaimed wood is well-seasoned, naturally lowering the likelihood of cracking or warping when compared to standard solid wood. Accordingly, it normally has a longer lifespan for finished products.

- Aesthetic and historical significance: Reclaimed wood often features distinctive characteristics, such as weathered textures, knots, and patina, that add a unique and rustic charm to finished building products. Particularly, reclaimed wood may be recollected from old buildings, factories, or other structures, adding various and unique stories to buildings. Therefore, it can enhance the visual appeal and historical characteristics of furniture, flooring, and other building parts.
- Support for sustainable practices: The demand for reclaimed wood encourages sustainable forestry practices. It promotes responsible wood sourcing and highlights the importance of utilizing existing resources. Furthermore, reclaimed wood requires less energy input, leading to a reduced overall carbon footprint in the production of wood-based products, as the process of harvesting, processing, and transporting new wood involves significant energy consumption.
- Cost-efficiency benefits: While initial costs for reclaimed wood may vary, its durability and longevity often result in cost savings over time. The need for frequent replacements is reduced, making it a financially viable option in the long run.

6.3.5 Bamboo

Bamboo is considered one of the best eco-friendly building materials. Bamboo usually grows rapidly in temperate, tropical, or subtropical climates. Bamboo has the properties of high strength-to-weight ratio, growth rate, hardness, and comprehensive strength. Due to these unique properties, bamboo is a versatile and sustainable material that has gained popularity in the construction industry as a viable alternative to traditional building materials. For instance, it has been produced into bamboo lumber, plywood, flooring, mat boards, veneer, composite materials, reinforcement bars, matting and woven products, charcoal, and fibreboard.

Bamboo has been proven to be as strong as steel due to its cylindrical form. It is even more durable than concrete, brick, or wood. The lightweight material can make it easier to work with lift, move, and create. Moreover, the material may be formed into sheets and planks that can be cut and laminated for use as walls, floors, roofs, and many other things. In comparison to conventional building materials, the elasticity of this material gives it a unique dimension. It is inherently waterproof and doesn't frequently warp in hot or humid conditions. However, bamboo requires special treatment to resist insects and rot. A starch present in the untreated bamboo, on the other hand, substantially attracts insects and causes it to bloat and break when exposed to water.

Case study: 2023 Annual Congress of Golden Keys China, green building materials in industrial park

The industrial park is a key scene for the industrial sector to achieve the "dual carbon" goal in China. In 2019, ESQUEL Group, headquartered in Hong Kong, built a textile industrial park named *Integral* in Guilin, Guangxi, China, which integrates Chinese landscape culture with the latest industrial concepts. Various measures from the construction to the operation phase of the park are aimed at reducing energy emissions. For example, building materials are reused on-site for landscaping and walls (as shown in Figure 6.2); buildings incorporate green designs to enhance natural light; photovoltaic panels on the roofs generate an annual average electricity output of 3 million kilowatt-hours. In addition, the park introduced green-washing processes, rainwater collection and recycling systems, and smart energy management systems to significantly improve production efficiency while reducing energy consumption and achieving zero industrial wastewater discharge. Statistics show that the energy consumption value per unit of building area during the operation and maintenance period of the *Integral* industrial park is only 131.8 kWh/(m^2.a), which is more than 30% lower than the national average.

The semi-enclosed bamboo strips enhance ventilation. The five main buildings are situated in a depression within the park, allowing for an open layout that effectively promotes the penetration of the southwestern monsoon in summer. The bamboo strips on the façades increase wind speed through the "air corridor", creating a more comfortable experience. In winter, the restored mountain slopes act as a natural barrier, blocking and mitigating the harsh northeasterly monsoon.

Figure 6.2 Green building materials: bamboo in Integral industrial park.

Source: zgsjlm.cn

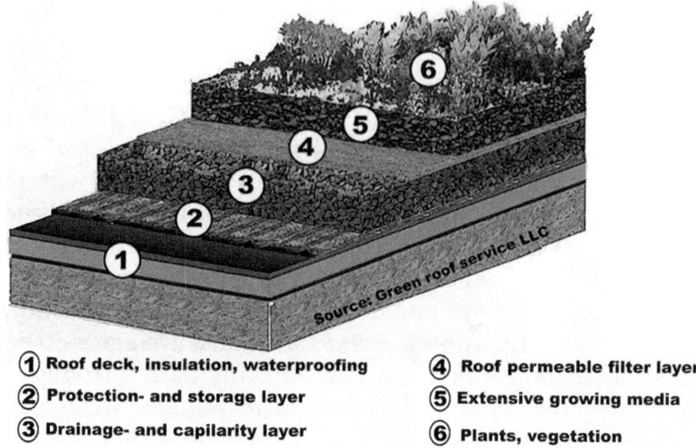

Figure 6.3 Functional layers of a typical extensive green roof.
Source: Archilovers

6.3.6 Green roof and wall

There are numerous examples of developing green roofs in many countries. A green roof is an effective energy-efficient way to reduce the building's cooling load in summer and heating load in winter. The green roof, also known as living roof or vegetated roof, is partially or completely covered with vegetation. A green roof is a succession of layers including a waterproof coating covered on a flat or slightly inclined roof so that it can retain stormwater, increase sound insulation, protect from fire, improve air quality, and keep warm inside. The roofing plants could be flowers, bushes, and other types of vegetation that flourish. Stormwater is more easily controlled when it is absorbed into the soil as opposed to a bare roof. Accordingly, the costs for heating and cooling of buildings are decreased, and the air quality is raised.

A green roof can be divided into three categories: extensive, intensive, and semi-intensive. The "extensive" green roof (as shown in Figure 6.3) is a basic type with lightweight and low-cost requirements, which could be retrofitted from existing buildings. The "intensive" green roof is a professional type that should be designed and constructed by professionals so that it can support heavier, deeper soil for intensive plants. The "semi-extensive" green roof is an intermediate type roof that is generally developed based on the roof condition by combining the features of both extensive and intensive green roofs. The loads of soil diversity and plant diversity and the building structure will affect the choices of green roof types. Therefore, green roof developers should consult a professional to check whether the building structure supports the green roof. The Australian Parliament House (as shown in Figure 6.4) in *Canberra* is probably one of the earliest and most successful green roofs all over the

Figure 6.4 Australian Parliament House green roof.

Source: Hydrosol

world, which is a kind of intensive green roof. The Parliament House building was constructed in 1988 for $1.1 billion.

Walls, fences, and other structures can also become homes for native plants and animals. There are varieties of green walls or "vertical garden" examples around us. The main types are where plants grow in planter boxes or a lightweight type of soil attached to a wall. Greening walls can insulate buildings, block out noise, beautify places, and benefit our health. However, the development of green walls necessitates an assessment of the wall's structural capacity to support the weight of the plants, soil, and water system, and therefore, consultation with a professional designer or engineer is required.

6.4 Recycling and reusing construction materials

Recycling materials has advantages in cost saving compared to buying new materials. The use of recycled materials is encouraged, particularly for achieving the sustainability objectives. Accordingly, local governments normally set restrictions on landfill waste and carbon emissions related to construction materials. Additionally, the government offers various freebies and other advantages for using recycled materials, such as financial subsidies (investment subsidy, production subsidy, and consumer subsidy), tax preference (e.g. value-added tax, corporate income tax, personal tax), and loan incentive (e.g. direct loan, credit loan, low-interest loan, and free interest loan). One of the most important parts of choosing recyclable building materials is to identify which types of materials can be recycled. Some building materials like concrete, steel, plastic, timber, and even bricks are recyclable. However, some plastics and broken glass cannot be recycled. The common waste materials are discussed in the following sections.

6.4.1 Recycled metal

Due to the benefits and requirements of sustainable development, recycling metal has become a common practice worldwide, as it can be recycled repeatedly without degrading its properties. Steel is the most recycled metal, followed by non-ferrous metals such as copper, stainless steel, and aluminium. Typically, recycled metal is first separated using magnets and other technologies, then processed with industrial shredders, and finally melted and purified into metal products for transport. In the construction industry, recycled metal is extensively used in both commercial and residential buildings, including reinforced concrete and steel structure parts. Compared to metal derived from mined ore, the benefits of using recycled metal include reduced consumption of energy, water, and natural resources, as well as decreased transportation and emissions. Additionally, recycled metal is often less expensive. For example, the façade of the Hard Rock Café in Bangkok is constructed using recycled aluminium, which has been recast into solid modules (as shown in Figure 6.5).

Figure 6.5 Façade of Hard Rock Café in Bangkok constructed using recycled metal.
Sources: Architectkidd

6.4.2 Recycled fibres

Recycled fibres are made by reprocessing fibres from the reclaimed fibres through the means of manufacturing processes and then making it into the final product or the component for incorporation into the product. Some commonly recycled fibres are nylon, wool, polyester, cotton, and down. In addition, as fibre waste has led to environmental concerns due to the disposal of landfills in the past decades, using recycled fibres has the benefits of environmental protection.

Fibres that are commonly used in the construction field consist of glass fibres, steel fibres, synthetic fibres (such as acrylic fibres and carbon fibres), and natural fibres (e.g. bast fibres and core fibres). In the construction field, recycled fibres are often applied in the formulation of lightweight concrete, due to the produced concrete with low density, high thermal insulation, and self-compacting abilities. Fibre-reinforced concretes have a wide range of applications. For example, glass fibres are beneficial for precast panels, curtain wall facings, sewer pipes, thin concrete shell roofs, wall plaster, and concrete blocks. Recently, carbon fibres have received growing interest due to their high tensile strength, special thermal characteristics, and so on.

The application of carbon fibre has recently attracted interest in various industrial fields such as aerospace, transportation, and sporting goods due to its high mechanical strength, corrosion resistance, high modulus, low density, and decreasing cost. Approximately 35,000 tons of carbon fibres were consumed in 2008, and this figure doubled in 2014 following a 12% annual growth (Pimenta & Pinho, 2011). However, only approximately 60% of virgin carbon fibre composite production is efficiently consumed in actual engineering applications, meaning that 40% of them are considered waste (Holmes IV, 2017). Given the limited space for landfilling and the high cost of disposing of this carbon fibre waste, the reuse and recycling of the increasing amount of carbon fibre waste become an issue that has important environmental and economic implications for sustainable development. Carbon fibre waste can enhance concrete mechanical performance in compressive strength, tensile strength, and modulus of elasticity (de Souza Abreu et al., 2020). Therefore, the carbon fibre composite waste from other industries can be considered as a kind of valuable resource, which is important in closing the loop of the carbon fibre life cycle for achieving sustainable development.

6.4.3 Recycled concrete aggregates

Aggregates are versatile materials used in various projects, including the manufacturing of concrete, asphalt, and drainage systems. Natural aggregates consist of sand, gravel, and crushed stone, while recycled aggregates are primarily made from recycled crushed concrete, bricks, and asphalt pavement. There is a growing demand and interest in aggregates from recycled sources, such as

Figure 6.6 Prahran Hotel in Melbourne, Australia.

Source: Arch20

construction and demolition waste and industrial by-products. Using recycled concrete aggregates as a substitute for crushed stone aggregates is both economical and environmentally friendly. Therefore, exploring the use of recycled aggregates is crucial for a sustainable future as it provides an alternative to traditional natural concrete.

Recycled concrete aggregate (RCA) is the most popular demolition and construction waste material used as both coarse and fine aggregate. For example, the Prahran Hotel in Melbourne features a façade made of recycled concrete pipes, offering diners a unique view and connection to the street outside (see Figure 6.6). The process involves breaking, removing, and crushing existing concrete into a material of a specific size and quality. The quality of recycled concrete aggregates, such as porosity and mechanical resistance, directly impacts the characteristics of new concrete mixes. This is because aggregates constitute 70% to 80% of concrete, directly affecting its overall quality.

6.4.4 Recycled glass

Glass is a commonplace material that is simply a translucent frozen liquid made of silica, soda ash, and calcium carbonate ($CaCO_3$) that has been liquefied at high temperatures. Glass has the properties of weather resistant to stand up rain, snow, sleet, and wind. It is also naturally corrosion-resistant and a high durability material like metals. Glass can be used for canopies, skylights, and roof partitions, even beams and flooring can be constructed out of it. And sometimes, entire façades made of glass are possible. It may be quickly recycled as a feedstock in the production of glass because it comes from a plentiful supply of glass waste.

Glass can be recycled indefinitely without losing its quality or worth. Recycled waste glass in concrete is often used as a partial replacement for at least one or more of its ingredients. Researchers investigated the use of waste glass as a partial replacement for fine or coarse aggregate together. The mechanical strength can be improved when adding recycled glass aggregate or powder in the construction field. However, it has been proved that adding a mixture of 5% glass powder and 10% glass sand combination to concrete will optimize both strength and durability properties in the production of eco-friendly concrete (Baikerikar et al., 2023).

6.4.5 Recycled asphalt

Asphalt is produced by the civil road building industry. Asphalt has the potential to be completely recyclable. In Australia, this level of recycling and usage of recycled materials in pavements has not yet been fully realized, although industry efforts are being made to change. Asphalt pavements on average are 4% bitumen and 96% aggregate. The top layer of asphalt, referred to as the "wearing course" (which is typically between 25 and 40 mm), is often removed and re-laid every ten to 15 years. A milling machine is used for removing the worn course. In order to ensure the physical properties of the mixture, the proper ratio of bitumen to aggregate, the correct proportion of aggregate size, and air voids, the recovered material is often transferred to an asphalt factory for sorting and bleaching.

6.4.6 Recycled plastic

Plastic waste is non-biodegradable and can persist in the ecosystem for many years. Burning plastic waste releases harmful compounds, including black carbon, into the air, soil, and water, endangering human health, animal welfare, and vegetation, while also contributing to global warming and air pollution. However, plastic waste can be repurposed as a sustainable material in polymers, concrete, lightweight concrete geo-pavements, reinforced geo-pavements, asphalt, and bricks. Almohana et al. (2022) reviewed plastic waste in producing sustainable concrete and pointed out that (1) plastic waste has low density compared to concrete aggregate, therefore, using it in concrete can reduce the unit weight of concrete, thus a reduction in the dead weight of the structure; (2) the use of plastic waste as a partial replacement (50% to 75%) for the total aggregate significantly boosts the efficiency of thermal and sound lightweight concrete insulation; (3) the cost of manufacturing plastic material is drastically diminishing in comparison to that of ordinary concrete, and plastic can be installed and utilized quickly with less labour due to its lightweight nature. Therefore, aiming for sustainable construction, plastic waste can be a typical material for the production of lightweight green concrete as a non-structural component in building construction.

Case study: Classroom of Hope rebuilt sustainable schools using recycled plastic blocks in Indonesia

Classroom of Hope has built sustainable schools that are safe and secure for communities in Indonesia, to help students return to their education quicker after earthquakes in 2018. The 300 m^2, five-classroom school building took just six days to construct by using 15 tons of plastic waste in the form of lightweight blocks (as indicated in Figure 6.7). The Block School was built with concrete foundations, lightweight blocks, steel rods, and U-profiles to reinforce the structure, and a galvanized steel roof, glass windows, and wooden doors to complete the building in 2021. The feedback from the local community has been very positive. The construction schedule is quicker compared to the common construction duration of three to six months for this type of school size, as the building blocks were fully interlocked in groups of two, four, and eight.

The revolutionary block solutions system produces plastic waste into lightweight, interconnecting plastic bricks that build earthquake-resistant buildings for Asia's first recycled plastic "eco-block" school. The building blocks are created by the Finnish company Block Solutions, which created the building material as an affordable solution for building low-income homes across Africa and Asia—all from recycled materials. The blocks are produced

Figure 6.7 An innovative new technology using recycled plastic blocks is rebuilding schools and homes in earthquake-ravaged Indonesia.

Source: Rotary

using materials from 50% recycled plastic waste and 50% leftover wood fibres from Finland's forestry industry. Moreover, in order to save supply chain costs from Finland, Asia's first Block Solutions factory has been contracted to be built to produce the blocks in Indonesia, so that Block Solutions' technology can be further applied to build public toilets, libraries, and administration buildings. It can be easily predicted that more building blocks will be produced, and more plastic waste will be recycled and reused.

6.4.7 Reused materials

Reused materials help save resources and energy by lowering the need for new construction materials. The reused materials include the following:

- Easy to remove items such as doors, hardware, appliances, and fixtures.
- Wood cut-offs can be used for cripples, lintels, and blocking to eliminate the need to cut full-length lumber. Scrap wood can be chipped on-site and used as mulch or groundcover.
- Infrastructure timber, such as power poles and railway sleepers, is another important source of reused hardwood, and there is a high demand for it in landscaping applications.
- Brick, concrete, and masonry can be recycled on-site as fill, sub-base material, or driveway bedding.
- Excess insulation from exterior walls can be used in interior walls as noise-deadening material.
- Paint can be remixed and used in garage or storage areas or as primer coats in other jobs.
- Packaging materials can be returned to suppliers for reuse.

6.5 Waste materials used in concrete construction

Concrete is considered the most frequently used construction building material with more than 10 billion tons produced each year in the construction industry, only second to water consumption. Cement, as one of the most vital contents of concrete, can cause considerable emissions of greenhouse gases like CO_2, which are a major factor in climate change. Besides, most of the waste materials are corrosive, hazardous, chemically reactive, and are frequently dumped in landfills, which pollutes the environment and causes health hazards. Therefore, a good sustainable construction strategy is required to utilize waste materials to reduce concrete aggregates in the sustainable construction industry.

6.5.1 Carbon fibre waste

Concrete with recycled carbon fibre can increase the porosity, compressive strength, workability, corrosion resistance, electrical conductivity, and microstructural properties. The splitting tensile and flexural strength can be improved for concrete after adding recycled carbon fibres. It has been concluded that the tensile strength of concrete can increase by 28.71%, 110.05%, 120.57%, and 188.52%, when adding with 0.25%, 0.50%, 0.75%, and 1% recycled carbon fibres, respectively. However, the tensile strength of concrete would be decreased if the percentage of additional carbon fibre was higher than 1% due to the increase in porosity (Akbar & Liew, 2020). Moreover, carbon fibres can also be used as the anode material for cathodic protection in reinforced concrete structures, which can reduce the risk of damage to the structures using lower current density.

Recycled carbon fibres can be used in concrete for the conductive concrete. Conductive concrete has the potential to be used in many different applications. The electrically conductive concrete can be used in underfloor heating which can provide a long-term, low-maintenance alternative to plumbed hot water installations. Additionally, the electric conductive concrete can be used for non-destructive testing and early damage by detecting its stress and damage in the structural health monitoring systems.

Carbon fibre-reinforced plastic has been widely used in civil engineering for repairing, and replace deteriorated or damaged structural elements such as bridge decks, concrete beams, columns, walls, and slabs.

6.5.2 Fly ash

Fly ash is one of the most often used pozzolanic materials worldwide. It is an abundant industrial waste product from coal burning. Fly ash is a lightweight and fine residue that is then transported by flue gases and gathered by an electrostatic precipitator. The average size of fly ash particles is 20 microns, which is comparable to the Portland cement particles. Fly ash replacing the cement in concrete can also reduce the CO_2 emission, as each ton of cement releases 0.87 tons of CO_2 into the atmosphere, and the cement production has increased to 4.4 billion metric tons in 2021. The fly ash aggregates exhibit a high absorption rate of 16%–24.8%, a specific gravity range of 1.3 to 1.47, and a bulk density range of 650 to 790 kg/m^3 (Bhatia, 2016). Accordingly, fly ash concrete has significant advantages, such as improving workability; reducing the water amount; reducing bleeding; reducing drying shrinkage; reducing thermal cracking; releasing low heat of hydration; reducing the permeability of the structures; increasing resistance to chemical attack; decreasing the construction cost; and increasing the compressive strength, split tensile strength, and durability of concrete by 14%. Therefore, fly ash can also be utilized to make bricks and road construction. The most important area of fly ash is concrete production as a cementitious material.

Fly ash has been used in the concrete industry for over 50 years. Its application in construction can help minimize environmental pollution by reducing the need to dispose of the earth's plentiful resources, thereby decreasing costs and environmental harm. Large amounts of untreated fly ash, when disposed

Table 6.1 Fly ash production in different countries (Ghazali et al., 2019)

Country	*Amount of production (million tons/year)*	*Country*	*Amount of production (million tons/year)*
India	112	Malaysia	6.8
China	100	Canada	6
USA	75	France	3
Germany	40	Denmark	2
UK	15	Italy	2
Australia	10	Netherlands	2

of, can cause pollution of the land, water, air, flora, and fauna (Akmal et al., 2017). Consequently, many countries produced a significant amount of fly ash, particularly the Asian Countries. For example, as shown in Table 6.1, India and China collectively account for over half of the world's total production.

6.5.3 Silica fume

Silica fume is one of the waste materials used in the construction field. Silica fume is a grey-and-white dust product that is produced during the manufacturing process of smelting ferrosilicon alloys or silicon metal. Silica fume primarily contains 85%–95% amorphous SiO_2. Silica fume particles are extremely small and composed primarily of pure silica in non-crystalline form. The disposal of the silica fume leads to pollution and hazards in air, water, and soil. Adding silica fume to concrete could improve the compressive strength and the durability of concrete due to the reduction of permeability and refined pore structure and the reduction of the diffusion of attractive ions such as calcium hydroxide and chloride contents. Therefore, silica fume can be used in high-strength concrete, fine aggregate, repaired concrete, and dams.

6.5.4 Plastic waste

Plastic is a fantastic material that has been repurposed into many other building materials. Researchers are creating concrete from ground-up waste and recycled plastics instead of sourcing, mining, and milling new components. The procedure lowers greenhouse gas emissions while reusing plastic waste, which would otherwise fill landfills, a useful new use. Recycled plastic is used in concrete casting. The practice can reduce the greenhouse gas emissions and pollution. Alyousef et al. (2021) have reviewed the potential values of using recycled plastic aggregate (PA) and rubber aggregate (RA) in cementitious materials and believe PA and RA are cost-effective and sustainable construction materials incorporated in concrete, particularly for contributing to the functional properties of cementitious composites in reducing thermal conductivity and improving sound absorption and electric resistivity, even though higher concentrations of these aggregates would cause a greater reduction of concrete in workability, density, and mechanical strength (e.g. compressive, split-tensile, and flexural strength). In summary, recycled plastic waste could be used as sustainable aggregate in cementitious materials in concrete structures, particularly for lightweight structures.

6.5.5 Case study: Target Zero Waste to Landfill Study Home

The programme was funded by the Victorian Government's Beyond Waste Fund, RMIT (RMIT University) University, and the HIA (Housing Industry Association). In the programme, finally, the builder slashed the housing construction waste sent to landfill by 99%, surpassing the original "Target Zero Waste to Landfill" reduction target of 80%. This project was completed in 2014.

Phase 1: Construction of baseline home

This phase was mainly for collaboration and data gathering. The baseline home was constructed. During the construction, the waste production process and volumes were measured and mapped from the site. The production process and waste data were independently audited to understand construction impacts, establish a waste audit protocol, and identify what sort of waste was generated. It was found that 50% of the waste was due to bricks and 25% of the waste was from roof tiles.

Phase 2: Development of waste avoidance strategies

This phase focuses on gathering data and data analysis to find the opportunities that are drawn from the pulled auspices of the design teams together with the construction teams. This phase consisted of the following steps: to develop workshops for the grant recipients as the facilitators to design avoidance strategies, review building materials and emerging technologies, assess environmental impact, identify industry-ready solutions, and evaluate market acceptability for avoidance strategies. Furthermore, after completing the construction of that baseline home, the strategies were to develop various solutions with a complete redesign where they designed the whole façade of the house differently using different materials. The strategies also used the completely prefabricated system off the site, and all the waste was in the factory.

Phase 3: Implementing avoidance strategies to redesign homes

These avoidance strategies include using metal roof pre-cut off-site, concrete bricks in lieu of clay bricks, square set plaster finish, brick window in-fills, and temp bracing at the frame stage for noggins; eliminating plasterboard cornice; and working with the supplier to deliver an exact number of polystyrene pods required for slab. The key findings of the Target Zero Waste to Landfill programme will help for the construction of demonstration homes to achieve the implementation of agreed avoidance strategies, onsite spoil separation and bulk bags for recycling, quantification of waste streams and volumes, waste stream and volume mapped from the site, and independent auditing of process and data. The final result indicated that 99% of the construction waste was diverted from landfills compared to baseline homes, and 72% of the total construction waste diverted from landfills was achieved with avoidance strategies.

6.6 Conclusion

This chapter examined the integration of sustainable building materials into construction practices, contrasting traditional materials with their green alternatives. It explored the benefits of using eco-friendly materials such as green concrete, thermal insulation, structural insulated panels, recycled wood, bamboo, and green roof and wall systems. These materials contribute to a smaller carbon footprint and enhance the energy efficiency and overall sustainability of buildings.

Additionally, the chapter discussed incorporating common recycled and reused construction materials to conserve natural resources and reduce waste. These materials include recycled metal, fibres, concrete aggregates, glass, asphalt, and plastic. Specifically, it highlighted the use of waste materials in concrete construction to promote sustainability and reduce environmental impact, such as carbon fibre waste, fly ash, silica fume, and plastic waste.

Promoting green building materials and recycled and reused materials offers numerous environmental benefits, including protecting biodiversity and ecosystems, improving air and water quality, reducing waste streams, and conserving natural resources. Further R&D in sustainable construction materials is expected to expand the markets for green products and services and promote sustainable construction practices.

References

Akbar, A., & Liew, K. J. J. O. C. P. (2020). Assessing recycling potential of carbon fibre reinforced plastic waste in the production of eco-efficient cement-based materials. *Journal of Cleaner Production*, *274*, 123001.

Akmal, A. M. N., Muthusamy, K., Yahaya, F. M., Hanafi, H. M., & Azzimah, Z. N. (2017). Utilization of fly ash as partial sand replacement in oil palm shell lightweight aggregate concrete. *IOP Conference Series: Materials Science and Engineering*, 012003.

Almohana, A. I., Abdulwahid, M. Y., Galobardes, I., Mushtaq, J., & Almojil, S. F. (2022). Producing sustainable concrete with plastic waste: A review. *Environmental Challenges*, 100626.

Alyousef, R., Ahmad, W., Ahmad, A., Aslam, F., Joyklad, P., & Alabduljabbar, H. (2021). Potential use of recycled plastic and rubber aggregate in cementitious materials for sustainable construction: A review. *Journal of Cleaner Production*, *329*, 129736.

Baikerikar, A., Mudalgi, S., & Ram, V. V. (2023). Utilization of waste glass powder and waste glass sand in the production of Eco-Friendly concrete. *Construction and Building Materials*, *377*, 131078.

Bhatia, A. J. J. S. T. A. (2016). Greener concrete using coal industrial waste (fly ash): An efficient, cost-effective and eco-friendly construction materials for the future. *Journal of Civil Engineering and Construction Technology*, *1*, 8–14.

Chen, L., Huang, L., Hua, J., Chen, Z., Wei, L., Osman, A. I., Fawzy, S., Rooney, D. W., Dong, L., & Yap, P. S. (2023). Green construction for low-carbon cities: A review. *Environmental Chemistry Letters*, *21*(3), 1627–1657.

De Souza Abreu, F., Ribeiro, C. C., Da Silva Pinto, J. D., Nsumbu, T. M., & Buono, V. T. L. (2020). Influence of adding discontinuous and dispersed carbon fibre waste on concrete performance. *Journal of Cleaner Production*, *273*, 122920.

Ghazali, N., Muthusamy, K., & Wan Ahmad, S. (2019). Utilization of fly ash in construction. *In IOP Conference Series: Materials Science and Engineering*, *601*(1), 012023.

Holmes IV, J. C. (2017). *Composite materials and related methods for manufacturing composite materials.* U.S. Patent Application 15/350,976.

Pimenta, S., & Pinho, S. T. (2011). Recycling carbon fibre reinforced polymers for structural applications: Technology review and market outlook. *Waste Management*, *31*, 378–392.

Index

For Product Safety Concerns and Information please contact our EU representative GPSR@taylorandfrancis.com Taylor & Francis Verlag GmbH, Kaufingerstraße 24, 80331 München, Germany

Batch number: 10397794

Printed by Printforce, the Netherlands